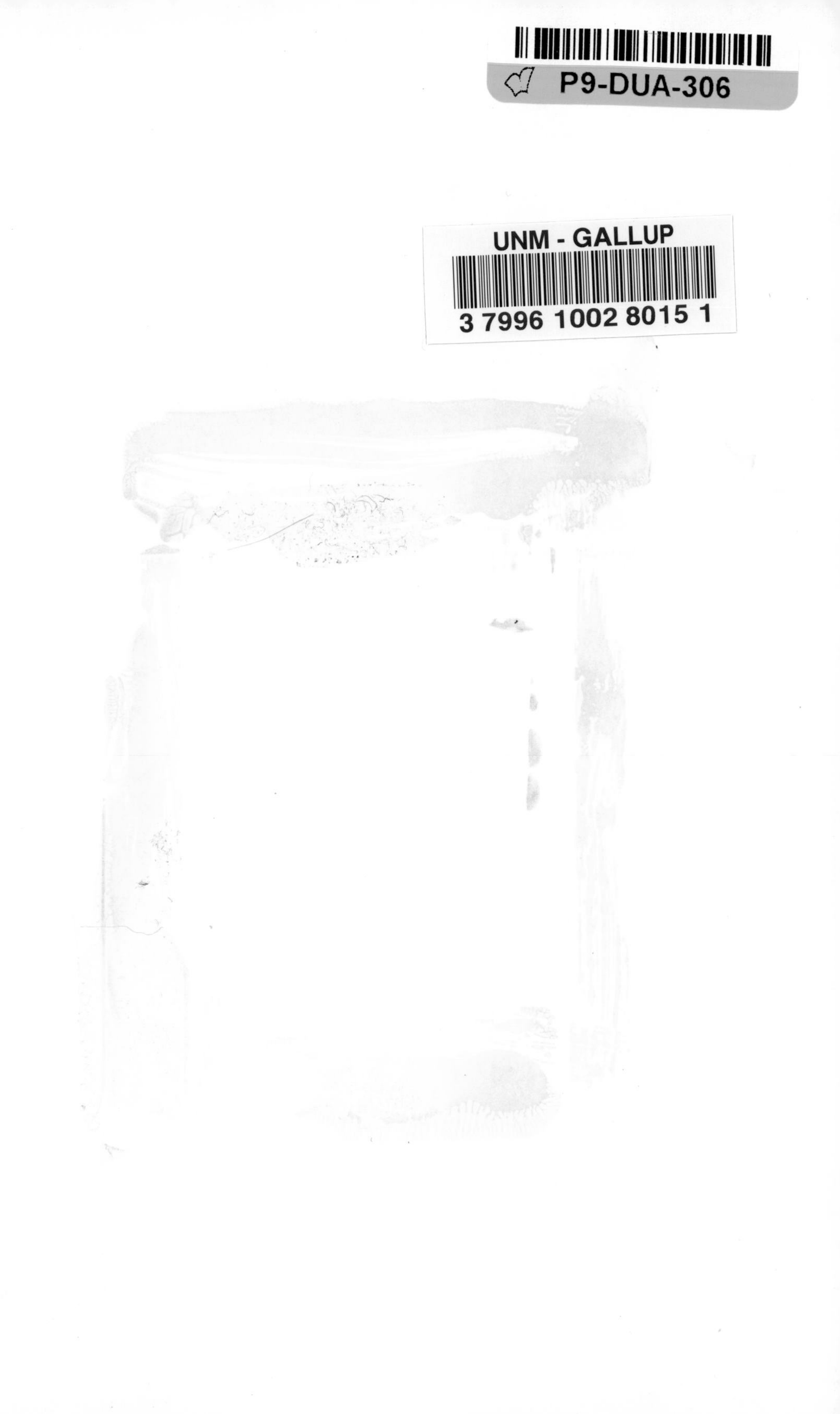

90 Years and 535 Miles

90 Years and 535 Miles

Vegetation Changes Along the Mexican Border

ROBERT R. HUMPHREY

UNIVERSITY OF NEW MEXICO PRESS
Albuquerque

Library of Congress Cataloging in Publication Data

Humphrey, Robert R. (Robert Regester), 1904–
90 years and 535 miles.

Bibliography: p.
Includes index.
1. Vegetation dynamics—Mexican-American Border Region—History. 2. Botany—Mexican-American Border Region—Ecology—History. 3. Phytogeography—Mexican-American Border Region—History. 4. Mexican-American Border Region—History. 5. Vegetation dynamics—Mexican-American Border Region—History—Pictorial works. 6. Botany—Mexican-American Border Region—Ecology—History—Pictorial works. 7. Phytogeography—Mexican-American Border Region—History—Pictorial works. 8. Mexican-American Border Region—History—Pictorial works. I. Title. II. Title: Ninety years and five hundred thirty-five miles.
QK142.H86 1987 581.97 86-30896
ISBN 0-8263-0945-3

First edition

Contents

To Roberta—wife, friend, associate,
and many things besides,
whose constant support, companionship, and help
have made this study pleasurable and possible.

90 Years and 535 Miles

The U.S.-Mexican border, like all others,
has no intrinsic existence. It was established
piecemeal as a result of warfare, compromise,
exigency and occasionally, common sense.

—Paul Ehrlich, Loy Bilderback
and Anne Ehrlich
in *The Golden Door*

Introduction

Scientists have long accepted the fact that the kinds and numbers of plants and animals inhabiting any given area of this planet are in a constant state of flux. These changes may be seasonal or annual, or they may occur, or only be apparent, over long periods. They may be temporary or they may appear to be more or less permanent. The kinds and degrees of these changes, as well as a determination of whether assumed changes have, in fact, occurred, must have intrigued man since the beginning of his ability to reason.

The initial determination of observed changes leads to the question: Why? Or, at times, the knowledge that there has been a change in the environment resulting from abnormal climatic phenomena, volcanic eruptions, or man's ever-increasing influence, may stimulate a desire to determine the impact of these factors on the affected area and on its plant and animal life.

The present study was suggested by a series of photographs of the monuments that designate the U.S.-Mexico boundary between the Rio Grande and the Colorado River. Taken in 1892 and 1893, these pictures provide some evidence of the more conspicuous vegetation then common at each monument location. The exact location of each monument and consequent photo could be accurately determined, and I felt that by taking repeat pictures from the original camera situations I might be able to determine the vegetational and erosional changes that had occurred during the intervening years. For the most part I did not expect to be able to distinguish specific kinds of grasses or to determine grass densities with a high degree of accuracy. Many shrubs and trees, on the other hand, I felt could be identified. At very least, life-form changes would be clearly evident in a study of the early photos. The photographs would be supplemented by early written accounts and by my own field record of today's vegetation and erosion.

Two hundred fifty-eight of these monuments were erected during the three-year period from 1892 to 1894 to mark that portion of the international boundary lying west of El Paso, Texas.

I decided to carry the study only across New Mexico and Arizona, omitting California. This would necessitate obtaining repeat photos of the 205 monuments from El Paso, Texas, to the Colorado River, a total distance of 535 miles. Because the boundary crosses several mountain ranges and two extensive sand dune areas, I anticipated that the study would be physically demanding to both man and machine. This proved to be the case. Roads adjacent to or near the border are often nonexistent, or, if shown on maps as present, they were often impassable or, in some instances, no longer existed.

The best maps usually available were the U.S. Geological Survey Topographic Quadrangle sheets. These varied in usefulness, however, since many had not been updated recently, in the most extreme instance since 1917 (sixty-seven years).

Travel and living quarters were provided by a 1978 Volkswagen Camper. A four-wheel-drive vehicle would have been more useful in some instances; in others it would not. As we set up camp wherever the end of the day found us, it would not have provided the conveniences and comfort of the VW. And besides, one often uses what one has.

Fieldwork was started in May 1983 and finished in November 1984. Where roads existed adjacent to the border, access to the monuments was comparatively easy. Often, however, these roads required considerable repair work and brush cutting to make them passable. Where there were no roads we would drive as close to the border as possible, then I would take off on foot or, in a few instances, on horseback. As the monuments were located on the highest mountain ridges wherever the boundary crossed a mountain range, the walking at times became difficult or, at times, hazardous. I had no wish to be packed out on a stretcher because of a broken leg, or worse. Eventually, however, I was able to rephotograph all except two of the original monuments.

Historical Background

The boundary between the United States and Mexico was initially established by the Treaty of Guadalupe Hidalgo in 1848, then modified in 1853 under the Gadsden Treaty to define the border as it exists today. This final agreement, however, was not the result of an amicable afternoon spent over a cup of tea or even a few bottles of Mexican beer. Instead, it was reached only after a war between the two nations that began in 1846 and ended in September 1847 when American troops conquered Mexico and even occupied Mexico City.

From the Gulf of Mexico to El Paso the center of the Rio Grande was defined as the dividing line between the two countries. Where there were multiple channels, the center of the deepest was designated as the boundary.

From El Paso the line continued westward under the Gadsden Treaty from the Rio Grande "where the parallel of 31°47′ north latitude crosses the same; thence due west one hundred miles; thence south to the parallel of 31°20′ to the 111th meridian of longitude west of Greenwich; thence in a straight line to a point on the Colorado river twenty English miles below the junction of the Gila and Colorado rivers; thence up the middle of the said river Colorado until it intersects the present line between the United States and Mexico." This *present line* had been agreed to February 2, 1848, under the Treaty of Guadalupe Hidalgo as: "a straight line drawn from the middle of the Rio Gila, where it unites with the Colorado, to a point on the coast of the Pacific ocean distant one marine league south of the southernmost point of the port of San Diego . . ." (Emory, 1857).

From the Rio Grande west to the Colorado River the boundary was marked originally by forty-seven monuments. Although these were on or close to the International Boundary, they were not located fixed distances apart but seem to have been built very often where fancy and an abundance of easily available rocks dictated. For the most part they were made of loosely piled mounds of stones, although a few at key locations were built of carefully constructed masonry. Some of these latter, for example, Nos. 1, 2, 3, 40, 46, and 53, have been maintained and are in use today. In California "a substantial monument" established the western end of the line; another was located at the junction of the Gila and Colorado rivers to indicate the eastern end of the California portion of the boundary. Between these extremities five others were built. The other forty-seven were variously located along what is today the Arizona–New Mexico boundary with Mexico.

In the ensuing years, as settlements and mining claims adjacent to the boundary developed, inevitable boundary and rights-of-access disputes arose until the decision was reached by both governments that the International Boundary would have to be marked by a series of immovable, unmistakable, more closely spaced monuments. As a consequence, representatives of the two governments met in Washington, D.C., in the summer of 1882 and agreed upon the details of locating and building the monuments that mark the boundary today.

There were conception-to-delivery delays typical of bureaucratic endeavors, but work was finally begun by February 1892. During the ensuing twenty-nine months 257 monuments were constructed in place of the original 54; 205 of these between the Rio Grande and Colorado River, 52 from the Colorado to the Pacific.

The survey of the boundary was run by Major William H. Emory during the period 1849–55.* His initial appointment was as chief astronomer and topographical engineer of the Boundary Commission. In this position he received little cooperation either from the Departments of Interior and War in Washington or from the boundary commissioner, John R. Bartlett. From the first, when Bartlett tried to force acceptance of an incorrect location for the initial monument on the Rio Grande, on which the entire line to the west would depend, until his final removal and replacement by Emory as commissioner, Bartlett's interests lay more in exploring the Southwest than in the less exciting responsibilities of his job as commissioner.

Emory finally received notice of his appointment to replace Bartlett as boundary commissioner in August 1854. Under Emory's capable direction, work progressed smoothly and efficiently, and he was able to report finally that the survey had been completed within the time prescribed by Washington and "at an expense much within the appropriation made by Congress." Note that there were no cost overruns or budgets out of balance!

Specifics of the new monuments that were to be built are given on page

*For the full story the reader is referred to Faulk (1967), Wagoner (1975), and the concise account by Bufkin (1983).

13, Part I of Senate Document 247, 55th Congress, 2nd Section (Senate Document 247, 1898a).*

> Where stone shall be found in sufficient abundance the monuments may be of stone, and in other localities shall be of iron, in the form of a simple tapering, four-sided shaft with pediment, rising above the ground to a height of 6 feet and bearing suitable inscriptions on its sides. These monuments shall be at least two centimeters in thickness and weigh not less than 500 pounds each.

As most of the monuments were ultimately made of iron, rather than stone, and as the original specifications were rigidly adhered to and describe these monuments as they stand today, I quote the full specifications here, as given on page 18, Part I of Senate Document 247.

> The monuments shall be cast either of steel or iron, as may be found most desirable;
>
> The height 6 feet and thickness of metal 2 centimeters, as provided in the treaty;
>
> The size at the base to be 12 inches square, and at top 9 inches square;
>
> The pyramid at top to have a base 9 inches square and six inches in height;
>
> The base of the monument will have a flange 4 centimeters in thickness, where it joins the sides, diminishing to 2 centimeters at the outer edge, and will be 6 inches wide; through this flange on each side will be a hole 1 inch in diameter; these holes are to receive the fastening bolts to secure the monument in place;
>
> The monument will be filled either with concrete or sand well packed, as may be found most practicable, and to be fastened in place by four 1-inch bolts well secured to the natural rock foundation, where possible. Where no rock foundation is available a foundation of concrete, 3 feet square and 2 feet deep, will be prepared of Portland cement and sand in proportion of about one to three. In these cases the fastening bolts will extend through the foundation and be secured below by suitable heads and washers; the upper ends of these bolts will be provided with a thread and nut, the latter round and put on with pipe tongs. . .
>
> On the west side of each monument may be attached two socket rings for holding a flagstaff, for the purpose of more easily verifying the

*Senate Document 247 (The U.S. Boundary Commission Report) consists of two volumes. The first of these contains the text of the document; the second is an album of photographs without text.

line between monuments remote from each other; these rings will be placed, one at the top of shaft, the other 12 inches below, and will be carefully located in the plane of the boundary;. . . .

The weight of the monuments as designed is 710 pounds, which is 200 pounds greater than the weight suggested in the treaty.

The estimated cost of the monuments seems so low, when compared with today's costs, as to verge on the ludicrous. For example, quoting from page 18, Part I of Senate Document 247:

Estimate of cost of monuments, including concrete bases, transportation and setting in place:

Castings, 710 pounds at 4 cents	$ 28.40
Wrought bolts and nuts 86 pounds at 5 cents	4.30
Concrete, 18 cubic feet	15.00
Sockets for flagstaff	1.00
Transportation by wagon	15.00
Transportation by pack mules	10.00
Setting in place	15.00
Contingencies	11.30
Total for each monument	$100.00

After the work was begun it was found that, due to the "exigencies of transportation" this cost estimate was too low, so the figure was, almost apologetically, increased to $150.00 per monument. What might today's cost be—$1500, $2000, or even more?

The initial plan called for casting each monument as a unit that would then be carried by either wagon or mule-back to its location for erection. This soon proved not to be feasible, perhaps because a 710-pound package carried on one side of a mule was a physical (or biological) impossibility, while two such weights, one on each side for balance, was too much of a load for even the best Missouri or Mexican mule. Consequently, each monument that had to be packed in by mule was cast initially in six sections; i.e., two side plates at 135 lbs. each, two side plates at 110 lbs. each, one base plate at 110 lbs., and one top section at 110 lbs., for a total of 710 pounds.

Soon thereafter, construction of the monuments was again modified slightly. As given on page 19, Part I of Senate Document 247:

After a number of sectional monuments had been put up, the design was changed to meet the difficulties of pack transportation. The new design provided 7 pieces—a base, a cap, and 5 intermediate sections, the latter each 14 inches high, resting one above another, and all held

> in place by a vertical bolt connecting the base and top section, the cap being fastened with rivets.
>
> The sectional monuments, when erected, were of the same appearance as those cast solid.

The 1882 agreement called for "a preliminary reconnaissance of the frontier line . . . within six months," with a view to determining the condition of the original monuments constructed following signing of the Gadsden Treaty. However, items of more immediate interest or greater pressure seem to have occupied Washington, with the result that work was not begun, much less completed, within the stipulated six-month period. Consequently, a second convention was called for 1889 that revised the provisions of the 1853 treaty. This time a more realistic expiration period of five years was agreed to. Construction of the monuments that mark the boundary today was initiated and carried to completion in accordance with the recommendations of this 1889 convention.

Senate Document 247, Part II, pages 8–14, includes a list by name of all the Americans involved in locating and setting up the monuments. The following paragraph from Part II, page 10 is of interest.

> The photographic work was first undertaken by J. H. Wright. He was soon found to be ill-adapted to field service, and his place was supplied by M. J. Lemmon for a short period. He was also unsatisfactory, and Mr. D. R. Payne was employed in August 1892, and continued to the close of the entire work, rendering most faithful and efficient service in his profession as photographer, and also as overseer in the erection of monuments.

Although the text does not say that Payne was unusually capable, his photographs attest to his photographic ability. In addition, I note that on August 6 he is cited as "photographer and general assistant." When next referred to in November of the same year, he was called "photographer and overseer"; then when again mentioned, presumably the following year, the emphasis on his responsibilities had changed and he was listed as "overseer and photographer."

The monuments were first photographed under conditions and using equipment which, although representing the best then available, by today's standards would seem almost insurmountably crude and difficult. For example (Senate Document 247, Part II, page 199):

> The camera was of the portable type, with 8 by 10 inch glass plates. . . . The necessary plate-holders, printing frames, baths, a supply of chemicals, and a developing tent were also provided. The outfit was usually transported in one of the spring wagons, but on portions of

> the line where wagons were impracticable Mr. Payne managed to carry the camera and other necessary articles on horse or mule back. The trails were frequently rough and sometimes dangerous, but happily no serious accident ever occurred to the camera or its belongings.

Emory was conscientious in administering the funds available for erecting the monuments to a degree which, in today's era of cost overruns, seems unrealistic (Sen. Doc. Part II, p. 206). In his final accounting, of the total $251,804.72 appropriated for the fieldwork, Emory showed an unexpended balance of $701.90. This was then applied to the office work, preparing maps, and writing reports. For this phase of the operation, he had a total of $20,702.66. Even this was not expended in full, and his final accounting showed a "Balance on hand at completion of final report" of $7,159.50. In addition, the 258 monuments from the Rio Grande at El Paso to the Pacific Ocean at San Diego were erected or constructed between February 1892 and May 1895, apparently within the allotted time span.

Biotic Areas

Traveling from east to west along the International Boundary from El Paso, Texas, on the Rio Grande to San Luis, Sonora, on the Colorado River one passes through five biotic communities. These, using the classification of Brown, Lowe, and Pase (1979) are: Chihuahuan Desertscrub, Semidesert Grassland, Madrean Evergreen Woodland,* Sonoran Savanna Grassland, and Sonoran Desertscrub.

The biotic regions traversed by the International Boundary have been described by various individuals and in a variety of contexts. For this reason it seems redundant here to do more than treat them synoptically as a means of orientation.

CHIHUAHUAN DESERT (Chihuahuan Desertscrub)

This is a low-shrub desert with a precipitation pattern in which the greater part of the year's moisture falls as summer thunderstorms. As pointed out by Shreve (1942), this is typically a high desert with elevations usually ranging from about 4,000 feet to well above 6,000. The bulk of the desert lies in Mexico in the states of Chihuahua, Coahuila, San Luis Potosí, Zacatecas, and Durango. In the United States it extends primarily into southwestern Texas and southern New Mexico. Lesser areas also occur in southeastern Arizona.

The perennial vegetation of the Chihuahuan Desert is today largely made up of low-growing shrubs intermixed with occasional low-growing cacti. The occasional trees are restricted for the most part to infrequent riparian habitats, as are most of the grasses. As a general rule, however, this desert contains few trees, no tall cacti, and few perennial grasses.

* "Madrean" from the Sierra Madre of Mexico where this community centers and is most widely represented.

Biotic Areas

As is true of any extensive biotic region, the Chihuahuan Desert comprises many ecosystems and a consequent wide variety of plants. That portion traversed by the boundary contains only a few of these systems, and only the vegetation contained in these is pertinent to this analysis and will be included here. Although many perennial taxa were recorded, the most characteristic were honey mesquite, creosotebush, mariola, tarbush, and ocotillo. An authoritative discussion of this and the other North American deserts is given by Forrest Shreve in his excellent publication *The Desert Vegetation of North America* (1942). Subsequent accounts will be found in Jaeger's *The North American Deserts* (1957) and in Brown's treatment of biotic communities of the American Southwest (1982).

Although shrubs typify most of that portion of my transect that traverses the Chihuahuan Desert, it crosses occasional swales of riparian communities where a drastic increase in soil moisture permits the establishment of coarse-growing grasses such as tobosagrass or sacaton. In one instance, at Monuments 29–31, an extensive tobosagrass swale had been plowed and the area converted to cropland. In other locations, areas that supported grasses in the 1890s no longer do today and appear to represent an invasion of Chihuahuan Desert scrub into former semidesert grassland.

Semidesert Grassland

The biotic region referred to here as the *semidesert grassland* was originally described and mapped as *desert-grassland transition* by Shreve (1917). Since then it has been variously designated and is perhaps best known as the *desert grassland* (Humphrey, 1958). For a more complete synonymy, see Brown (1982, p. 123).

As I have pointed out earlier (Humphrey, 1958, p. 4), this is usually not an essentially pure grassland prairie but typically consists of a mixture of grasses, forbs, and woody species. Extensive areas have been invaded by low-growing trees, principally mesquite, or by various shrubs, half-shrubs, or cacti, a process that continues into the present. This is a region in flux, capable of growing either grasses or woody species, or a mixture of the two. The temporary winner in this unceasing battle is determined by the pressure of factors that favor one or the other life form. Climate and soil generally favor the woody species, mesquite in particular, and, unless controlled in some way, they will eventually dominate the landscape.

Although many grass species typify the semidesert grassland, several of the gramas usually dominate. These most commonly include blue grama, sideoats grama, slender grama, sprucetop grama, and black grama. Other typical grasses may be curly mesquite, threeawns, beardgrasses, and tanglehead, among others. Velvet mesquite, prickly pears, and arborescent cholla-type cacti tend to be the predominant woody species. Two half-shrubs, burroweed and snakeweed, are often abundant and usually indicate depletion of the perennial grasses resulting from overgrazing or mesquite competition.

The semidesert grassland has been extensively studied in an attempt to determine the changes that have, and are, taking place and the underlying reasons for this evolution. A change of climate, the prevalence and/or control of fires, and grazing by livestock are the three factors usually considered largely responsible for a gradual takeover by the scrub species. Which of these is of paramount importance is often a moot question, and the one to which greatest importance is attached frequently depends on a particular research worker's major interest and field of expertise. This may result in an inadvertent lack of objectivity.

Evergreen Woodland

The region that I have here designated *evergreen woodland* was classified by Shreve (1915) as *Encinal* from the Spanish for *encina* or oak and the ending *al,* signifying *a stand of.* Hastings and Turner (1965) have used the term *oak woodland.* Their brief analysis of the terminology and description of this life zone well portrays its salient characteristics. Brown, Lowe and Pase (1979) have termed it *Madrean Evergreen Woodland.*

Shreve subdivides this life zone into an *Upper* and *Lower Encinal,* but I have not distinguished between the two here, this detailed a breakdown not facilitating my study.

That portion of the evergreen woodland traversed by the International Boundary lies in general above the semidesert grassland at elevations between 4,000 and 6,700 feet. The higher elevations, corresponding to Shreve's Upper Encinal, tend to have shrubby chapparal species such as Mexican manzanita and silktassel bush accompanying the trees that give the zone its chief character. These are usually Mexican blue oak, Emory oak, and alligatorbark juniper.

Because of the higher elevation and usual proximity to even higher mountainous terrain, precipitation here is higher and temperatures somewhat lower than in the adjacent semidesert grassland. The consequent more favorable moisture balance not only promotes establishment and growth of trees and shrubs, it also facilitates establishment of perennial grasses from the semidesert grassland below as well as a number that are particularly adapted to local conditions. This latter group commonly includes such coarse-growing species as mountain muhly, deergrass, and bullgrass.

An occasional, but still characteristic feature of the evergreen woodland is the major drainages where the flow is consistent enough to support a riparian community of distinctive trees and other vegetation. Although these are sometimes known as gallery forests, the term *forest* seems somewhat inappropriate here as applied to a stringer that is usually only a few yards or meters across but may be as much as a mile or even several miles in length. In any event, these riparian areas are chiefly characterized by a stream through steep, rocky banks with an intermittent or periodic flow during, and for a time after, the rainy seasons, and a tree overstory that is usually dominated by Arizona sycamore, velvet ash, Fremont cottonwood, Arizona walnut, and western hackberry.

SONORAN DESERT (Sonoran Desertscrub)

The Sonoran Desert has many faces, both spatial and temporal. The spatial aspects are best expressed by the perennial vegetation that characterizes the desert's various ecosystems. Seasonal changes in the ephemeral flora are short-lived, often to the point of evanescence, but the relatively long-lived perennials provide a durable taxonomic base for designating distinctive biotic differences or ecosystems. For this reason, the ephemerals, even though they may sometimes be a conspicuous component of the flora, I have usually mentioned casually or not at all. Instead, the perennial, usually woody, species are stressed. Most long-time vegetational changes will be reflected in these relatively long-lived taxa.

The characteristics and biota of the Sonoran Desert have been studied and written about by so many as almost to discourage any attempt to list them. Of all its students and researchers, though, Forrest Shreve is preeminent. His ecological treatment of this desert is classic as is the taxonomic component by Ira Wiggins in their monumental work *Vegetation and Flora of the Sonoran Desert* (Shreve and Wiggins, 1964). Other, more popular descriptions include those by Jaeger (1957), Kirk (1973), Helms (1980), Humphrey (1981), and Brown (1982).

The International Boundary transect of the present study intersects only a few of the many ecosystems that make up the Sonoran Desert, with a resultant comparatively small number of characteristic dominant species. The bulk of these are represented by low-growing trees or shrubs, perennial grasses occurring only occasionally or not at all. As might be expected, the greatest variety of grasses occurs in the eastern portion of this desert where it adjoins the semidesert grassland. A few of the higher, more inaccessible mountainous areas also tend to have grass relicts. Finally, a few extremely xeromorphic grasses occur in the hot and arid western extremity of the transect. A few low-stature trees tend to prevail over most of the 190 transect miles of Sonoran Desert, most commonly velvet mesquite, foothill or blue paloverde, ironwood, and sahuaro cactus. Among the shrubs, various cacti of the genus *Opuntia*, creosotebush, triangle and white bursage, ocotillo, tomatillo, and saltbushes are usually the most common.

Except for Monument No. 1, which lies in what was originally a riparian cottonwood community on the west side of the Rio Grande, the first ninety-nine miles west from the Rio Grande and the thirty-one-mile dip to the south from Monument No. 40 to about six miles west of No. 53, lie within the Chihuahuan Desert and are characterized by various biotic communities of the Chihuahuan Desert. An additional sixty-three miles of this desert lie in Arizona.

Four plant communities predominate in New Mexico along this portion of the boundary: mesquite-palmilla, mesquite-mixed scrub, high-plain scrub, and tobosagrass. Except for about the first three miles (Monuments 1–3) the first fifty miles are characterized by low, active dunes of soft sand partially stabilized by clumps of low-growing honey mesquite interspersed with palmilla. West of the sand-dune area, the mesquite bottomlands occur periodically for short distances in, and adjacent to, drainages, but the bulk of the remainder is typified by the high-plains scrub community. Tobosagrass is encountered occasionally, particularly in swales toward the western end of this desert as one approaches the semidesert grassland.

Semidesert grassland extends from near Monument No. 56 westward for about twenty-four miles to Monument 65 on the crest of the San Luis Mountains where it is interrupted for about two miles by evergreen woodland. West of the San Luis Mountains the grassland continues across Animas Valley for about eleven miles to where an increase in altitude in the Guadalupe Mountains near the New Mexico–Arizona boundary results in a second incursion of evergreen woodland.

Dropping into Arizona to the west, the woodland gives way first to a brief stretch of riparian woodland in Guadalupe Canyon. This grades abruptly into the Arizona portion of the Chihuahuan Desertscrub that continues for about sixty-three miles westward past the towns of Douglas and Naco to the San Pedro River valley. Here the semidesert grassland resumes for about six miles across rolling plains and foothills to the southern end of the Huachuca Mountains.

Continuing westward from the Huachucas is an intermittent combination of evergreen woodland and semidesert grassland that stretches across the San Rafael valley and the Santa Cruz River, through the Canelo Hills and the Patagonia Mountains to Nogales. This admixture continues across the Pajarito Mountains, through the border town of Naco, and the southern end of the Baboquivari Mountains to about the eastern limits of the Tohono O'odham Indian Reservation.

At about the eastern boundary of the Tohono O'odham Reservation the vegetation changes to the more arid Sonoran desertscrub. This continues for approximately 140 miles across the reservation, and an additional 69 miles across increasingly arid and rugged terrain to the Colorado River.

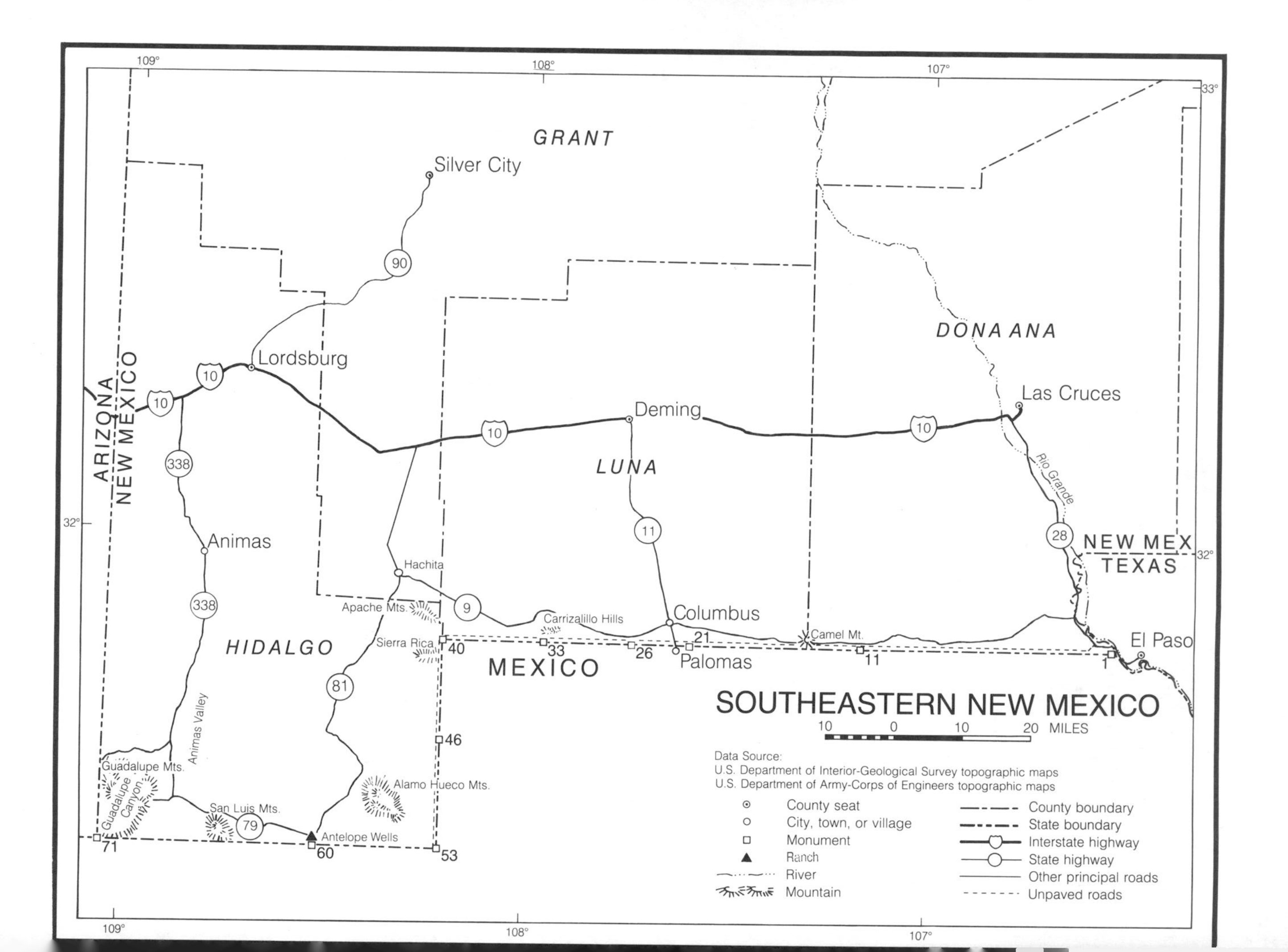

SOUTHEASTERN NEW MEXICO
10 0 10 20 MILES
Data Source:
U.S. Department of Interior-Geological Survey topographic maps
U.S. Department of Army-Corps of Engineers topographic maps
County seat
City, town, or village
Monument
Ranch
River
Mountain
County boundary
State boundary
Interstate highway
State highway
Other principal roads
Unpaved roads
GRANT
DONA ANA
LUNA
HIDALGO
ARIZONA
NEW MEXICO
NEW MEX
TEXAS
MEXICO
Silver City
Lordsburg
Deming
Las Cruces
Animas
Hachita
Columbus
Palomas
El Paso
Rio Grande
Camel Mt.
Carrizalillo Hills
Apache Mts.
Sierra Rica
Animas Valley
Guadalupe Mts.
Guadalupe Canyon
San Luis Mts.
Alamo Hueco Mts.
Antelope Wells
10
90
338
11
9
28
81
79
71
60
53
46
40
33
26
21
11
1
109°
108°
107°
33°
32°

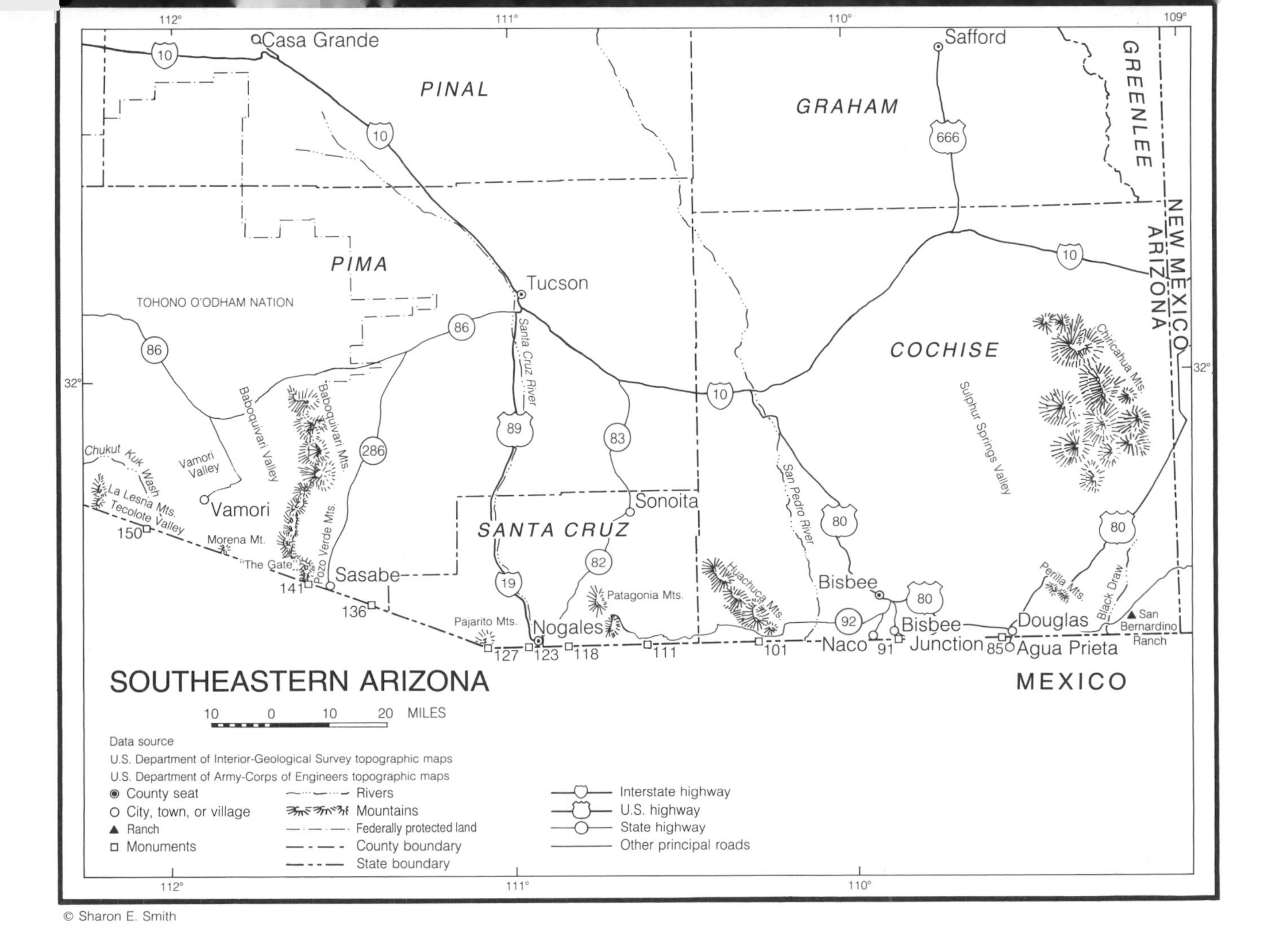
SOUTHEASTERN ARIZONA
NEW MEXICO
ARIZONA
MEXICO
GREENLEE
GRAHAM
COCHISE
PINAL
PIMA
SANTA CRUZ
TOHONO O'ODHAM NATION
Safford
Casa Grande
Tucson
Sonoita
Nogales
Sasabe
Vamori
Bisbee
Bisbee Junction
Naco
Douglas
Agua Prieta
San Bernardino Ranch
Chiricahua Mts.
Perilla Mts.
Black Draw
Sulphur Springs Valley
San Pedro River
Huachuca Mts.
Patagonia Mts.
Santa Cruz River
Pajarito Mts.
Baboquivari Mts.
Pozo Verde Mts.
Baboquivari Valley
"The Gate"
Morena Mt.
Vamori Valley
La Lesna Mts.
Tecolote Valley
Chukut Kuk Wash
10
666
80
92
83
82
89
19
86
286
85
91
101
111
118
123
127
136
141
150
109°
110°
111°
112°
32°
10 0 10 20 MILES
Data source
U.S. Department of Interior-Geological Survey topographic maps
U.S. Department of Army-Corps of Engineers topographic maps
County seat
City, town, or village
Ranch
Monuments
Rivers
Mountains
Federally protected land
County boundary
State boundary
Interstate highway
U.S. highway
State highway
Other principal roads

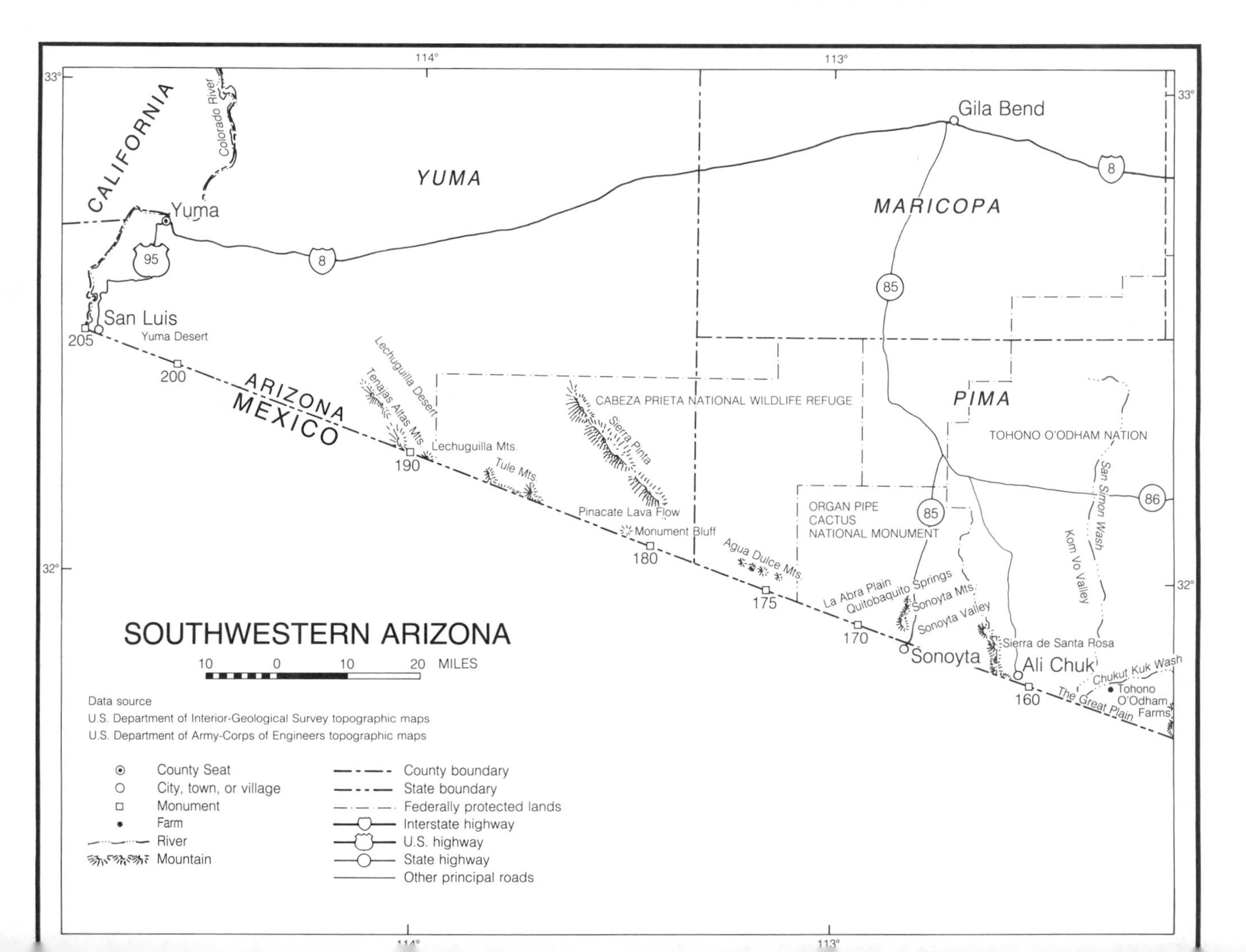
SOUTHWESTERN ARIZONA
10 0 10 20 MILES
Data source
U.S. Department of Interior-Geological Survey topographic maps
U.S. Department of Army-Corps of Engineers topographic maps
County Seat
City, town, or village
Monument
Farm
River
Mountain
County boundary
State boundary
Federally protected lands
Interstate highway
U.S. highway
State highway
Other principal roads
CALIFORNIA
Colorado River
Yuma
YUMA
MARICOPA
PIMA
Gila Bend
San Luis
Yuma Desert
205
200
190
180
175
170
160
ARIZONA
MEXICO
Lechuguilla Desert
Tenajas Altas Mts.
Lechuguilla Mts.
Tule Mts.
Sierra Pinta
CABEZA PRIETA NATIONAL WILDLIFE REFUGE
Pinacate Lava Flow
Monument Bluff
Agua Dulce Mts.
ORGAN PIPE
CACTUS
NATIONAL MONUMENT
TOHONO O'ODHAM NATION
San Simon Wash
Kom Vo Valley
La Abra Plain
Quitobaquito Springs
Sonoyta Mts.
Sonoyta Valley
Sonoyta
Sierra de Santa Rosa
Ali Chuk
Chukut Kuk Wash
Tohono
O'Odham
Farms
The Great Plain
8
95
85
86
114°
113°
33°
32°

Then and Now 1892 to 1984

In the discussion that follows, plant cover, erosion, and other pertinent items are discussed for each monument area in sequence from Monument No. 1 at El Paso, Texas, westward to No. 205 on the Colorado River.

As the original photographs were taken in 1892 and 1893, the recent ones either in 1983 or 1984, the time lapse between the two series may vary by one or, at most, two years. Thus, a ninety-one- to ninety-two-year period separates the earlier and later pictures.

Plant common names are used in the text, but each of these is listed alphabetically in the Appendix with the appropriate scientific name.

Three possible methods of classifying the vegetation were considered: (a) by a count of taxon density; (b) by measurement of ground area covered by each taxon; or (c) by general appearance of the area, i.e., the most readily apparent or principal life forms. Although either of the first two of these would have yielded data most capable of future repeat evaluation, the large number of sites involved, the fact that these kinds of data were not available from the original material sources, and the time available for the study prevented their use. As a consequence, the third, or general appearance method, was employed.

Each monument is designated by number and by the biotic community in which it falls. For example, No. 2—Chihuahuan Desert; No. 65—Evergreen Woodland; No. 66—Semidesert Grassland; No. 143—Sonoran Desert, and so on.

The decision to use the current appearance of each area as the criterion for classification obviated the need to base the classification on the potential or climax vegetation an area might support. Thus, an area that was grassland in 1892 but was growing woody plants in 1984 has been classified as scrub rather than grassland.

Descriptions of the Monuments

MONUMENT NO. 1. Urban

This, the easternmost monument on parallel 31°47′, was built in 1855. After some initial delays and correspondence between the Mexican and United States boundary commissioners, José Salazar y Larregui and William H. Emory, they and nine other representatives of the two governments met on the west bank of the Rio Grande at the agreed-upon site for the first monument, and the foundation was laid January 31, 1855.

A paper, written in both Spanish and English, and signed by both commissioners and their attendants, was placed in a bottle and buried five feet deep beneath the foundation. The paper read:

> We, the undersigned, have this day assembled to witness the laying of the foundation of the monument which is to mark the initial point of the boundary between the United States and the Republic of Mexico, agreed upon, under the treaty of Mexico, on the part of the United States by William Hemsley Emory, and on the part of the Republic of Mexico by José Salazar y Larregui, latitude 31°47′ (Emory, 1857, Vol. I, p. 28).

The monument of cut stone was twelve feet high, five feet square at the base, and tapered to two feet six inches square below a capstone with a sharp taper to a point at the top. The lower four feet of the cement base were jacketed with a thick layer of cement mortar in 1892, and the original stones are now almost obscured by a plaster coating; otherwise it remains recognizable as the original.

In 1855 the monument was recorded as being 71.04 meters west of the center of the Rio Grande. By 1892, when construction of the line of monuments to the west was initiated, the river channel had shifted 101.50 meters to the east, leaving the monument 172.6 meters west of the new center. Shifting river channels do not make for reliable boundaries.

The photographs, taken facing to the northeast, show the riparian Rio Grande cottonwoods and associated species in the immediate background. The foreground in the original picture has been disturbed and shows no vegetation.

In the 1984 photo, the riparian species still show. And, in addition, several plants of yucca, agave, and mulberry have been planted as ornamentals in the small park that surrounds the monument. Aside from the floodplain cottonwoods, the excessive disturbance and introduction of exotics invalidate any possible use of these pictures for determining either vegetational or erosional changes here.

Monument No. 1. Urban

Monument No. 2. Chihuahuan Desert

This monument, like No. 1, is one of the original boundary markers that dates back to 1855. It has stood the test of time well and, although repainted occasionally, still looks much as it did 129 years ago, and certainly as it did when photographed 92 years ago.

The monument is located in the shallow rocky soil of the Sierra de Cristo as the mountains are known today or the Mulero Mountains as they were called in 1892. These rise precipitously, immediately west of the Rio Grande Valley. The change in name apparently comes from a large cross that has been built on a rocky ridge of the mountains and is visible from the monument. An annual pilgrimage is made to the cross on Easter Sunday.

The mountains here are steep and rugged, and we were strongly advised by the U.S. Border Patrol not to approach or enter the area without protection of some sort. It is apparently frequented by drug runners and Mexican bandidos. I saw no one when I took the repeat photo but was accompanied by a cooperative, armed border patrolman.

The only apparent vegetational change during the last ninety-one years seems to be an increase in the number and ground cover of woody vegetation. Today's species consist largely of creosotebush (estimated 95 percent), the remaining 5 percent being mariola and ocotillo.

Monument No. 2. Chihuahuan Desert

MONUMENT NO. 3. Chihuahuan Desert

Monument No. 3 lies on a shallow-soil caliche flat near the western edge of the Sierra de Cristo. This, again, is one of the original 1855 monuments and, like Nos. 1 and 2, has withstood well the weathering of time. In 1892 it was described as in good condition, "pyramidal in shape, 5 feet square at base, and built of rubble masonry, plastered on the surface with lime mortar. . . . They [Nos. 2, 3] were without inscription, but their height, 12 feet, and white color make them very conspicuous as boundary marks" (Sen. Doc. 247, p. 175).

Creosotebush makes up most of the sparse plant cover here today, with lesser amounts of crucillo, palmilla, ocotillo, and snakeweed. As was the case with the preceding monument, the only apparent vegetational change since 1892 has been a moderate increase in the background scrub cover.

MONUMENT NO. 3. Chihuahuan Desert

Monument Nos. 4–10. Chihuahuan Desert

The approximate forty miles of the boundary westward from the Sierra Cristo that these monuments include consist of low, hummocky sand dunes partially stabilized by a scattered growth of mesquite and a few other shrubs. The fine, loose sand makes driving over most of this distance in a two-wheel-drive vehicle impossible, and we were delighted when the U.S. Border Patrol in El Paso offered to provide a four-wheel-drive van and driver for a day while I did my fieldwork. With this invaluable assistance a portion of the border that had greatly concerned me was covered in one day with relative ease.

The soft sand and lack of water created extreme difficulties for both men and animals along this portion of the boundary while the monuments were being erected. Even modern vehicles with their large, pneumatic tires have a problem here with the fine sand, but so did the horse-drawn vehicles used in 1892. As given in Senate Document 247 Part II, page 181:

> In moving out from El Paso the monument party encountered great difficulties, owing to the heavy loads to be carried over the soft, yielding sands of the desert. For 50 miles west of the river no water was found near the line, and a supply had to be sent out in advance, together with several monuments and casks of cement. This labor tasked [sic] the limited transportation . . . to its utmost capacity, the strain on the animals being excessive.

Except for the information revealed by the photographs, few data are available on the former vegetation of this portion of the boundary.

In a general description of the boundary as far west as Monument 15, fifty miles from the Rio Grande, Mearns (1907, pp. 9, 10) says that this portion of the line traverses a dry, but productive desert. The eastern half, he says, "is covered with sand hills, built up by the low mesquite, sagebrush, and yucca." His *low mesquite* is the honey mesquite that is the dominant shrub there today; his *sagebrush* may be threadleaf sagebrush, the only member of the genus I recorded there in 1984, but that I found only occasionally. The *yucca* he refers to, as the early photographs verify, is the palmilla common there today. Mearns also lists a final species growing in the sand dunes, slender-branched dalea, that he mentions as "a very characteristic shrub of this desert." I observed none of this species in my study. However, as I was not attempting to develop a complete checklist of the boundary vegetation but was recording primarily those plants growing near the monuments, this dalea may still occur there. I did not, on the other hand, see any in the vicinity of the boundary markers.

The western half of the first fifty miles of the boundary, according to Mearns, is "covered with a black grama grass (*Bouteloua eriopoda* Torrey)." "Here and there" also were "large patches of the gregarious creosote bush" and crucifixionthorn, Mormon tea, and ocotillo. Desert holly "grew abundantly in the shelter of shrubs." He noted that "cacti were not numerous, though several kinds were found sparsely, and prickly pears were common in a few places." In general these same observations might be made today, with one major exception—no black grama or, indeed, grasses of any sorts, were encountered until we reached Monument 14, 47.5 miles west of Monument No. 1. There was a little black grama at No. 14, but none at either of the next two monuments or for many thereafter.

As Mearns was careful to list black grama by both its common and scientific names, there should be little doubt that his identification was correct. It must be assumed, therefore, that there has been a marked decrease in the prevalence of black grama along this portion of the boundary during the interval since the monuments were erected. Note here that in my later discussion of Monuments 6 and 7, Senate Document 247 mentions grama grass at these specific locations, although no perennial grasses grow in the immediate vicinity of these monuments today.

Observations relating specifically to Monuments 4–10 follow.

Monument No. 4

Monument No. 4. Except for palmilla and possibly honey mesquite, it is impossible to identify with certainty any of the species in the 1892 photo. Palmilla and honey mesquite are still dominant in the area today, and there have been no apparent changes in either species composition or plant density.

Monument No. 4. Chihuahuan Desert

Monument No. 5. The quality of the earlier picture is poor, making plant identification difficult. There does, however, appear to have been a better vegetation cover in 1892 than there is today. Honey mesquite shows in the earlier photo and was probably the dominant species. Western honey mesquite is also present today (showing in the extreme right side of the recent photo), and probably occurred in 1892 as well. The shrub breaking the skyline on the far left of the 1892 photo is one of these two mesquite varieties. A single ocotillo shows to the right of the monument in the earlier photo; none occurred in the general area in 1984. As ocotillos tend to be rather short-lived, this does not necessarily indicate a long-term vegetational change.

The original picture indicates no apparent wind erosion; the later one reveals a wind-blown sandy terrain only partially stabilized by scattered clumps of honey mesquite and palmilla. These same species show in the earlier photo but are interspersed with a rather good ground cover of other vegetation.

Monument No. 5. Chihuahuan Desert

Monument No. 6. Although the quality of the 1892 photo is poor, making plant identification difficult, it does show a strikingly better plant cover than exists there today. As was the case at the preceding monument, the earlier photo indicates no apparent wind erosion; the later one reveals a wind-blown sandy terrain only partially stabilized by scattered clumps of honey mesquite and palmilla.

Senate Document 247, page 182, mentions grass at this monument and at No. 7: "The sites of 6 and 7 were in all respects similar to those for 4 and 5, except that grama grass began to appear in sufficient quantities to afford some grazing for the animals. . . ." Whether this consisted only of species of grama or, in part, of some other grass, such as tobosa, must remain problematical. Because tobosagrass begins to appear today at about this point along the boundary, it is possible that this species, or galletagrass, in addition to the black grama mentioned by Mearns, also occurred here. There can be no doubt, however, that perennial grasses were once abundant here.

MONUMENT NO. 6. Chihuahuan Desert

Monument No. 7

Monument No. 7. The 1892 photo shows a ground cover in the vicinity of Monument No. 7 that was considerably heavier than when rephotographed in 1984. Both honey mesquite and palmilla show as probable dominants in the earlier photo, as they definitely were in the later one. However, the 1892 pictures shows a scattered, but good ground cover of grass; there is none there today. In its place is a sparse stand of snakeweed.

No wind erosion is apparent in the earlier photo; in contrast, the recent picture shows active erosion, with sand that is only partially stabilized by the large clumps of mesquite.

MONUMENT NO. 7. Chihuahuan Desert

Monument No. 8

Monument No. 8. Monument No. 8 is located on the exposed crest of a low hill of actively blowing sand. Both the earlier and the later pictures show this erosion, also that the hilltop supports little vegetation. Honey mesquite, which is evident in the 1892 photo, is abundant there today. The fine-stemmed half-shrub to the left of the monument in the recent photo is turpentine broom, as probably are the similar appearing plants in the earlier photo.

A single creosotebush shows on the extreme right of the recent picture and is the principal shrub visible in the background of both photos. It appears to be more abundant today throughout the background than formerly.

Monument No. 8. Chihuahuan Desert

Monument No. 9. Vegetational or erosional differences between the earlier and later photos are slight, if any. Wind erosion, which was active when the 1892 picture was taken, is still so today. Although the ocotillos so prominent in the earlier picture have died, there are others nearby that do not show in the repeat photograph.

Today's vegetation in the immediate vicinity includes creosotebush, Mormon tea, ocotillo, palmilla, and snakeweed. All of these except Mormon tea and snakeweed can also be identified in the earlier photo.

MONUMENT NO. 9. Chihuahuan Desert

Monument No. 10. Wind erosion in the vicinity of Monument 10 has been active and extensive since 1892. At that time there was some sand movement; however, during the ninety-two intervening years the sand mounds have grown and the depressions between have deepened. Note that the accumulation of sand in the recent photo half obscures the monument that was almost entirely visible ninety-two years earlier.

Today's vegetation consists largely of honey mesquite, palmilla, and snakeweed, with the mesquite a strong dominant. These same species appear to have predominated in 1892.

Monument No. 10. Chihuahuan Desert

MONUMENT NO. 11. Chihuahuan Desert

The region of active dune formation lies largely east of Monument No. 11 today, as it did ninety-two years ago. The soil adjacent to the monument had been disturbed by construction activities and was bare of vegetation in the earlier photo. The background, in contrast, and in contrast to the situation today, appears to have supported a good stand of tobosagrass. Today no tobosagrass remains, having been replaced largely by snakeweed. Other occasional woody species that occur there today are honey mesquite (seen to the immediate right of the monument) and Mormon tea.

MONUMENT NO. 11. Chihuahuan Desert

Monument No. 12. Chihuahuan Desert

As with the preceding monument, there has been a marked change here over the years in plant life form. The once abundant tobosagrass has been replaced by snakeweed. Some Mormon tea may have been present earlier but is not identifiable in the photograph; it is rather common today. Scattered clumps of honey mesquite are discernible in the earlier picture as they still are today.

Wind erosion was slight before; it is much more evident today.

MONUMENT NO. 12. Chihuahuan Desert

MONUMENT NO. 13. Chihuahuan Desert

Monument 13 has vanished and only the overturned cement base remains. My repeat photo shows Border Patrolman Joe Brewster standing on the base with Camel Mountain in the background.

The vegetation today appears to be essentially the same as in 1892, consisting primarily of honey mesquite, creosotebush, palmilla, tarbush, mariola, and snakeweed.

The ground cover is sparser today than formerly, permitting more soil and sand movement by the frequent winds. Although one grass species, tobosagrass, was sufficiently common in 1891 to be visible and identifiable at Monuments 7, 11, 12, and 13, I recorded none in 1984 east of Monument 14, 47.5 miles west of the Rio Grande. Although there may formerly have been a few more grass species in this region, the blowing sand provides a poor habitat for establishment of most grasses, and any that might once have been there may have been killed by the livestock that still graze the area.

MONUMENT NO. 13. Chihuahuan Desert

Monument No. 14. Chihuahuan Desert

Monument 14 is located on the north-facing slope of a rocky, shallow-soil hillside. The site shows no apparent change either in plant cover or erosion during the ninety-two-year period that has elapsed since the earlier picture was taken. The most common woody species growing in the vicinity of the monument today are mariola, fourwing saltbush, range ratany, and shrubby buckwheat. Two species of perennial grass, galleta-grass and black grama, are fairly common and may be assumed to have also been there when the monument was erected September 27 or 28, 1892.

Monument No. 14. Chihuahuan Desert

Monument No. 15. Chihuahuan Desert

Although the vegetation here is sparser today than when the monument was built, the woody species composition appears to have changed little through the years. Honey mesquite surrounds the monument today, and the plant that now partially obscures the monument may be assumed to be the same, but much smaller, individual that shows in about the same position in the 1892 photo. The other scrub species most common here today are fourwing saltbush, palmilla, creosotebush, and snakeweed. Tobosagrass, and perhaps other grasses, show in the original picture. In contrast, there are no grasses in the immediate vicinity today.

Although there may have been some wind erosion in 1892, there are now rather large accumulations of sand in the mesquite clumps, indicating that soil movement, which is currently moderately active, has been so for many years.

Monument No. 15. Chihuahuan Desert

Monument No. 16. Chihuahuan Desert

The two photographs show a marked change in plant composition during the last ninety-two years—a shift from grasses and a sparse stand of shrubs to an essentially pure stand of shrubs and halfshrubs. Tobosagrass or, less likely, galletagrass, was abundant in 1893; there is none there today. Traces of another grass that often indicates range deterioration, sand dropseed, were recorded in 1984.

The principal species in the area today are the suffrutescent mariola, snakeweed, and threadleaf sagebrush, with an occasional honey mesquite and creosotebush.

MONUMENT NO. 16. Chihuahuan Desert

Monument No. 17. Chihuahuan Desert

There seems to have been no major vegetational or erosional change in the vicinity of Monument 17 during the past ninety-two years. Large clumps of honey mesquite characterize the area today, as they did earlier. Fourwing saltbush and honey mesquite are the only shrubs I noted there in 1984. The mesquite is obvious in the earlier picture; saltbush, although possibly present, cannot be identified from the photo. Nor can any grasses, although a few weak plants of sacaton do occur there now; some show in the foreground of the recent photo.

Wind erosion and low dune formation are active today as they were in 1892.

Monument No. 17. Chihuahuan Desert

Monument No. 18. Chihuahuan Desert

The area in the vicinity of Monument 18 appears to have much the same vegetation today as it did ninety-one years ago. Honey mesquite is still dominant, including some plants which, if not the same individuals, are at least growing in the same places. As might be expected, they have grown somewhat taller. Because of the severe drought that had begun two years before and still prevailed in September of 1892, the mesquites were largely leafless, in marked contrast to their condition when I rephotographed them in 1984.

Two species of woody plants predominate here today: honey mesquite and fourwing saltbush. Snakeweed is abundant locally, although little shows in the picture. An extensive sacaton flat in the background extends almost the full width of the recent photo. Although the image is not clear, there is every reason to assume that the same stand of sacaton was present in the earlier picture. This same grass is rather common today near the monument, although none shows in the photograph.

Moderately active wind erosion is evident in both pictures.

Monument No. 18. Chihuahuan Desert

Monument No. 19. Chihuahuan Desert

Although the principal plant species here today are probably the same as in 1892, the honey mesquite on the right side of both pictures has grown considerably in both area and height. The bedraggled and almost leafless appearance of the shrubs in the earlier photo is probably due in part to overgrazing and in part to drought.

Shrubs and half-shrubs are dominant in the immediate area today, principally honey mesquite, fourwing saltbush, and snakeweed. Although there are no grasses close to the monument, the slight rise on which it is located lies in an extensive sacaton flat. This shows in the background of both pictures.

Despite the wind erosion that is still moderately active here, the shifting sands are more stable today than formerly, seemingly due in part to the fact, as indicated by the exposed monument base, that much of the original loose sand has been blown away.

Monument No. 19. Chihuahuan Desert

Monument No. 20. Chihuahuan Desert

Both plant cover and erosion seem to be little changed at Monument 20 today from their condition in 1892. Soil movement was essentially stabilized then, as it still is. The only visible change is in the cover of perennial grasses. The earlier picture shows a good stand of what appears to be tobosagrass throughout the background. Some of this remains, but as small, scattered plants. A few clumps of sacaton also show in the recent picture although much more is growing in adjacent lower-lying areas. As this is a long-lived species, it may be assumed that some of the grasses showing in the 1892 photo were also sacaton.

The same woody plants that occurred at the two preceding monuments—honey mesquite, fourwing saltbush, and snakeweed—are common here also. In addition, a little seepweed was observed, indicating a local area with poor drainage and high salt content.

This area, as well as the range for many miles to the west, has long been subjected to heavy grazing, and it is probably due largely to the poor forage quality of both tobosagrass and sacaton that these grasses are still there today.

Monument No. 20. Chihuahuan Desert

Monument No. 21. Chihuahuan Desert

The 1892 photo shows grass, probably largely tobosagrass, where none occurs today at this monument. Although the earlier photo shows a grass-scrub mixture, a comparison of the two pictures shows a marked increase of woody plants with the passage of time at the expense of the former grasses. This area, like that at Monument 21, has been heavily grazed for many years, a type and degree of land use that has favored the woody plants, many of which have little or no forage value.

The same scrub species are abundant here as to the east, namely honey mesquite, fourwing saltbush, palmilla, and snakeweed. This vegetation, as with the mixture of grasses and shrubs that formerly prevailed, has been dense enough to greatly reduce wind erosion.

MONUMENT NO. 21. Chihuahuan Desert

MONUMENT NOS. 22 AND 23. Chihuahuan Desert

Monument 22 lies on the eastern edge of the small Mexican border town of Palomas; No. 23 lies on the western edge. Because of urban pressures and livestock grazing, the vegetation of this entire area, particularly on the Mexican side of the line, has been greatly altered.

Monument 22 was photographed looking north into the United States, the 1984 picture showing a stand of sacaton, fourwing saltbush, and seepweed. Vegetation on the Mexican side of the line here today consists mostly of ephemerals, largely Russian thistle. The original photo shows what appears to be fourwing saltbush and honey mesquite. The absence of mesquite in the recent photo may be due to its eradication as a means of increasing the carrying capacity of the rangeland for domestic livestock.

At Monument 23, although honey mesquite occurs sparsely on both sides of the boundary, sacaton and tobosagrass are moderately abundant, intermixed with lesser amounts of honey mesquite and fourwing saltbush. On the Mexican side of the line, the same ephemerals occur as at Monument 22.

Monument No. 22. Chihuahuan Desert

Monument No. 23. Chihuahuan Desert

Monument Nos. 24–26. Chihuahuan Desert

When I photographed the monuments along this part of the boundary on May 9, 1984, a new dirt road, extending westward from Palomas, had recently been built close to and south of the fence, totally destroying the vegetation. South of this denuded strip the natural vegetation still remained. North of the fence, except for a narrow, ungraded ranch road, the vegetation had not been drastically modified.

The earlier photos of all three of these monuments show a fair cover of perennial grasses and a comparatively sparse stand of shrubs. In contrast, I recorded no perennial grasses at any of these monuments when they were rephotographed. At that time the shrubs consisted largely of creosotebush with lesser amounts of snakeweed. Creosotebush, although much less abundant in 1892, appears to have been intermixed with a few arborescent cacti, these two species appearing to have been almost the only shrubs growing there at that time.

The level to gently rolling terrain is little subject to erosion, and none shows in either set of photos.

Monument No. 24. Chihuahuan Desert

Monument No. 25. Chihuahuan Desert

MONUMENT NO. 26. Chihuahuan Desert

Monument No. 27. Semidesert Grassland

Monument 27 is located in a typical tobosagrass swale that appears to have changed little during the intervening ninety-two years. Although tobosagrass may also be found on upland areas, these swales are, or at one time were, vegetated by almost pure stands of tobosagrass. They occur typically in low-lying depressions of both the Chihuahuan and Sonoran deserts that are subject to occasional saturation or even flooding. Within these deserts, they might perhaps best be considered as islands or, more appropriate physiographically, as lakes in a desert matrix. In the present context, however, I am considering them as discrete units of semidesert grassland.

In addition to tobosagrass, this site also supports occasional plants of palmilla and Mormon tea, as it did ninety-two years earlier. Because this is an area of moisture accumulation and the soil is well protected by the grass, no erosion was apparent when the first picture was taken, nor is it today.

Monument No. 27. Semidesert Grassland

Monument No. 28. Chihuahuan Desert

The 1892 photo shows a tobosagrass, rock-studded swale with thinly scattered creosotebushes. Ninety-two years later the tobosagrass has gone, and an almost pure stand of creosotebush has taken over.

It is interesting to note here that creosotebush has replaced the former tobosagrass although it had not at the preceding monument. Slightly better drainage in the present instance may provide a habitat better suited to the creosotebush, giving the slight impetus needed to favor it under a grazing regime inimical to the grass.

Monument No. 28. Chihuahuan Desert

Monument Nos. 29 and 30. Chihuahuan Desert (Cropland)

The tobosagrass that formerly characterized this stretch of the border has now been eradicated by plowing. In the vicinity of Monument 29 the Mexican plowed fields had been abandoned when rephotographed in May 1984, and were growing only a crop of annual weeds (largely Russian thistle), burroweed, and snakeweed; those adjacent north of the line had just been replowed after the previous year's crop of cotton. At Monument 30 the land on both sides had been freshly plowed, the only vegetation being scattered roadside or ditchside plants of Russian thistle, snakeweed, and burroweed.

MONUMENT NO. 29. Chihuahuan Desert (Cropland)

Monument No. 30. Chihuahuan Desert (Cropland)

Monument No. 31. Chihuahuan Desert

Monument 31 stands prominently on an eastern outlier of the Carizalillo Hills, on the crest of a low rocky hill. The vegetation here seems to have changed little over the last ninety-two years. Although only 2.1 miles west of Monument 30 and 500 feet higher, the plant growth here is in marked contrast to the original tobosagrass of the two previous monuments. Instead of tobosa, the hills here support a mixture of shrubs and perennial grasses, the shrubs represented largely by scrub oak, little-leaf squawbush, algerita, mariola, fourwing saltbush, tomatillo, and beebrush; the grasses by sideoats grama, slender grama, and wolftail. All of these species are characteristic of upland, well-drained sites that have not been excessively overgrazed.

Despite the proximity of this site to the readily accessible flatlands below, it seems to have been protected to some extent from excessive overgrazing by the steep hillsides and rocky terrain. As a net result, there has been little to no discernible change in either plant composition or total ground cover since the earlier picture was taken.

Monument No. 31. Chihuahuan Desert

Monument No. 32. Chihuahuan Desert

Except for a few differences in species, the vegetation here is similar to that at Monument 31; both are located in similar terrain in the Carizalillo Hills. The earlier and later photos, supplemented by my field notes, indicate no basic change in plant life form since the monument was constructed.

The principal woody species at this site today are scrub oak, mariola, banana yucca, one-seed juniper, sotol, ocotillo, crucifixionthorn, algerita, and honey mesquite. The larger dark-colored woody plants left of the monument in the immediate background of the recent photo are largely juniper; the shrub in the lower right-hand corner is scrub oak.

Intermixed with the woody vegetation are a number of perennial grasses, most of which were grazed too closely to be identified with certainty. The principal grass seemed to be galletagrass, although this, too, was severely grazed and may have been the closely related tobosagrass. It may be assumed that there were also other grasses here, probably the same species as at Monument 31.

Monument No. 32. Chihuahuan Desert

MONUMENT No. 33. Chihuahuan Desert

There has been a marked increase of scrub vegetation in the vicinity of Monument 33 during the last ninety-two years. Although the earlier picture shows too little detail to indicate clearly that there was a grass cover where there are shrubs today, it may be assumed that this was the case. Palmilla, showing abundantly in 1892, occurs sparsely today. Palmilla is typically a grassland species and in level terrain of this sort indicates grassland. No perennial grasses occur in the immediate vicinity today. My field notes, recorded May 10, 1984, read: "There are no grasses here; the range is in very poor condition on both sides of the fence, and there probably was a large amount of grass at one time."

The principal scrub vegetation showing in the recent photo is honey mesquite, tarbush, Mormon tea, fourwing saltbush, snakeweed, and creosotebush. The low-growing shrub that fills the right foreground of the recent picture is mesquite.

Monument No. 33. Chihuahuan Desert

MONUMENT NO. 34. Chihuahuan Desert

The descriptions of vegetation contained in Senate Document 247 are few in number, sporadic, and not necessarily accurate. They do not always agree with the photographic evidence. For example, on page 183, Part II, the document reads: "A plain about 10 miles in width and abounding in a heavy growth of mesquite extends westward to the Apache Mountains. Along this part of the line were erected Monuments Nos. 34, 35, 36, and 37. . . ." As will be noted in the discussion of these monuments that follows, the early vegetation of No. 34 was largely creosotebush, at No. 35 it was grassland, No. 37 may have contained some mesquite but only No. 37 almost certainly had a "heavy growth" of this spiny shrub.

The two photographs of Monument 34 show no apparent change in the vegetation. An extensive tobosagrass flat may be seen in the far background of both pictures. The principal vegetation around the monument today consists largely of tarbush and creosotebush with lesser amount of mariola, snakeweed, desert zinnia, and desert holly. There are no perennial grasses.

Monument No. 34. Chihuahuan Desert

MONUMENT NO. 35. Chihuahuan Desert

The vegetation here has changed drastically during the last ninety-two years. What formerly appears to have been essentially a pure stand of tobosagrass has now become an open mixture of creosotebush and snakeweed, with no grass. Although this is a well-drained upland, and not a typical tobosagrass site, this grass seems to have been almost the only species here when the earlier picture was taken. The fact that the site is atypical may explain the disappearance of the grass from a region where it has persisted in nearby, lower-lying swales.

MONUMENT NO. 35. Chihuahuan Desert

MONUMENT NO. 36. Chihuahuan Desert

The 1892 photo shows a mixture of shrubs and grasses; ninety-two years later there are no grasses and the shrub cover has become more dense. Tarbush appears to be a dominant shrub in the earlier picture. No snakeweed is distinguishable and mesquite, if present, cannot be identified. There does seem to be a partial understory of grasses, probably tobosagrass. Today, the plant cover consists solely of scrub species, largely honey mesquite, creosotebush, tarbush, and snakeweed, with minor amounts of crucifixionthorn and desert zinnia.

MONUMENT NO. 36. Chihuahuan Desert

MONUMENT NO. 37. Chihuahuan Desert

The photographs do not indicate any apparent change in composition of the vegetation here. The earlier photo shows too little detail to determine whether or not grasses were present in 1892; when the repeat picture was taken in 1984, however, there were none. The two yuccas shown to the right of the monument in the earlier photo are no longer there, but others are growing nearby.

Scrub-type vegetation is dominant today, as it was earlier, the principal species being creosotebush as a strong dominant, with western honey mesquite, tarbush, mariola, banana yucca, and snakeweed secondary. A mesquite shows against the skyline to the right of the monument in the recent picture and was probably also a common species there earlier.

MONUMENT NO. 37. Chihuahuan Desert

Monument No. 38. Chihuahuan Desert

There has been little or no discernible change in the vegetation here during the time interval that has elapsed since the first photograph was taken. The monument is located on the extreme north end of the Sierra Rica, a low range of limestone hills that extends south into Mexico. The vegetation was, and still is, primarily scrub, with an understory of grasses. Although the grasses were too closely grazed at the time of the repeat photo to make certain identification possible, one species appeared to be galletagrass. Sacaton is locally abundant in adjacent low-lying areas and is probably the dominant species in the island-like light-color areas that show in the background of the earlier picture. Some of these may also have been vegetated by tobosagrass. Today these grasses have been replaced in large part by scrub vegetation.

The principal woody species adjacent to the monument today are ocotillo, creosotebush, feather dalea, mariola, little-leaf squawbush, and Palmer agave.

There is no comment in Senate Document 247 on the vegetation here, only the observation: "This ridge slopes to the northward, and its steep, rocky scarp rendered the work of hauling the monument to its position very trying on the animals and men" (p. 183).

MONUMENT NO. 38. Chihuahuan Desert

Monument No. 39. Chihuahuan Desert

Vegetational changes at this monument appear to have been minimal during the ninety-two-year interval. Scrub-type vegetation predominates in both photographs. The most abundant woody species occurring there today are creosotebush, feather dalea, kidneywood, ocotillo, Palmer agave, winterfat, and snakeweed, with lesser amounts of Mormon tea, little-leaf squawbush, and one-seed juniper. These are interspersed with a variety of grasses, principally sideoats grama, black grama, hairy grama, slim tridens, and large-flowered tridens. If there were any grasses when the 1892 photo was taken they are not distinguishable.

Identifiable woody species that show in the earlier photo are ocotillo, Palmer agave, one-seed juniper, and creosotebush.

MONUMENT NO. 39. Chihuahuan Desert

Monument No. 40. Chihuahuan Desert

Monument 40 marks the corner where the boundary along parallel 31°47′ makes a right-angle turn to the south along meridian 108°12′30″ for thirty-one miles before continuing westward along parallel 31°20′. As this constitutes one of the old original monuments built in 1855 by Emory and his men, a bit of its 1892 reconstruction history is of interest. Monument 40 "is the old cut-stone Monument No. 8 of the Emory map. The present measurements . . . show that this monument was originally located . . . too far east, thereby causing this section of the boundary to be 1 mile less in length than was provided by the treaty" (Senate Document 247 Part II, p. 183).

"Old Monument No. 8 . . . was taken down and rebuilt. . . . The same stones were replaced in their original positions, leaving the external appearance but little changed. Iron inscription plates were added, and also the new number, 40, to conform to the present series" (p. 186).

The vegetation at Monument 40 in 1892 was characterized then, as it is today, by various woody species. Ocotillo, Palmer agave, and one-seed juniper are identifiable in the early picture, also an apparently rather good stand of grass, probably tobosagrass.

One of Mearns's collecting stations was located at Monument 40, consequently he describes the area in some detail and lists "the most abundant shrubs and conspicuous plants" (Mearns, 1907). His list includes twenty-nine species, all except two of which were shrubs or small trees. He mentions no grasses, but this should not be taken to indicate their absence as his lists of the vegetation elsewhere, even where grasses must have been abundant, rarely include grass species. Even where he refers to grassy hills and plains he never mentions any grasses by name. It would appear that he was reasonably familiar with the woody vegetation and many of the forbs but possibly did not know the grasses and consequently made no effort to name them.

In contrast with Mearns's much more inclusive list of "most abundant shrubs and conspicuous plants," I recorded creosotebush, snakeweed, mariola, ocotillo, and western honey mesquite as the most abundant woody species. The only grasses near the monument were fluffgrass and slim tridens. There is none of the tobosagrass there today that seems to show in the earlier photo, although it does grow rankly in a drainage a short distance to the south and east.

MONUMENT NO. 40. Chihuahuan Desert

Monument No. 41. Chihuahuan Desert

Monument 41 was erected on a rocky hilltop, the highest point where the meridian line crosses the Sierra Rica. Too few plants show in the original picture for it to be of much value in determining possible vegetational changes. However, the excessive rockiness protects the plants from grazing animals, enabling many of even the more palatable species to survive.

Only ocotillo and Palmer agave can be distinguished in the earlier picture; both of these also occur there today.

Other species common near the monument today are creosotebush, feather dalea, mariola, sotol, and snakeweed. Although the general area is heavily grazed by cattle, the excessive rockiness tends to discourage livestock, with the result that several nutritious grasses, notably black grama, slender tridens, threeawn, and sideoats grama have survived.

Monument No. 41. Chihuahuan Desert

MONUMENT NO. 42. Chihuahuan Desert

Despite the sotol and ocotillo that show so prominently in the original picture, with none in the retake, there has been no general change in the vegetation of this area during the last ninety-two years. Scrub species characterized the area then, as they do today. If grasses were once present they do not show in the picture. None occurs there today.

The most abundant species today are creosotebush, tarbush, and gray coldenia, with a scattering of ocotillo, palmilla, and mariola. The shrub that is so conspicuous throughout the background of both pictures is creosotebush.

MONUMENT NO. 42. Chihuahuan Desert

Monument Nos. 43 and 44. Chihuahuan Desert

Originally a scrub-grassland mixture, this area has evolved into almost pure scrub. The grasses that show in the earlier photo of Monument 43 are probably largely either tobosa or sacaton. No grasses are to be found in the vicinity of either Monument 43 or 44 today.

The principal scrub species showing in the recent photos are creosotebush, tarbush, snakeweed, honey mesquite, western honey mesquite, and crucifixionthorn. With the exception of snakeweed, which has probably increased as the grasses disappeared, these same species may be assumed to have been dominant when the earlier pictures were taken.

Monument No. 43. Chihuahuan Desert

MONUMENT NO. 44. Chihuahuan Desert

MONUMENT NO. 45. Chihuahuan Desert

The upland with its shallow, rocky soils that characterized the sites at Monuments 43 and 44 has given way here to a poorly drained alluvium. The poor drainage, consequent moisture accumulation and the fine-texture soil provide a habitat suited to deep-rooted, tall, moisture-loving grasses. Thus, when the monument was erected this was, as it is today, a sacaton flat. A few shrubs show in the early photograph; a few more are growing there today. These are moisture-loving mesquites of both the honey and western honey varieties.

Monument No. 45. Chihuahuan Desert

Monument No. 46. Chihuahuan Desert

Although somewhat better drained than the area around Monument 45, this attractive monument is located in an alluvial area that does collect occasional runoff. The soil is dark in color, indicating a habitat where deep-rooted grasses once flourished.

The 1892 photo shows a rather open stand of scrub where today the picture and my field notes indicate "a veritable jungle." What may have been an essentially open sacaton flat at one time was being invaded even in 1892 by aggressive woody species that were competing with and replacing the grasses. By the time of my study in 1984, this invasion had largely ended, and the woody plants were in complete control.

Although scattered clumps of sacaton still remain, western honey mesquite and honey mesquite are now dominant with scattered plants of creosotebush, tarbush, and crucifixionthorn.

Monument No. 46. Chihuahuan Desert

Monument No. 47. Chihuahuan Desert

Monument 47 is located in a typical creosotebush flat that seems to have changed little over the years. Although creosotebush was dominant in 1892 as it is today, the earlier photo shows a low-growing ground cover that may have consisted either of perennial grasses or ephemerals or both. None of these were there when the repeat photo was taken. The abrupt difference in soil color at the fence line in the repeat photo is a result of excessive livestock trampling on the Mexican side of the line.

Although creosotebush is a strong dominant here today, there is an occasional mariola, tarbush, crucifixionthorn, and, particularly in the more disturbed areas, snakeweed. There are no perennial grasses.

About midway between Monuments 47 and 48 a ranch cross-fence heads westerly toward the Alamo Hueco Mountains. Beside the road here we found a rough slab of cement on which had been crudely traced in the once-wet cement the message:

> Frank Evans, born June 12, 1865, killed here May 1, 1907, by a crazy cook. Murdered in cold blood with an ax. Mark by Deacy, Aug. 6, 1947. Witness by Johnny Freman. Frank was good.

Monument No. 47. Chihuahuan Desert

Monument No. 48. Chihuahuan Desert

The mixture of grasses and shrubs that prevailed in 1892 has changed over the years to an almost pure stand of shrubs. The earlier photo shows an abundance of grass, probably tobosa, in an open scrub. Woody plants predominate today, with only an occasional clump of tobosagrass.

Most of the woody vegetation now consists of creosotebush, with tarbush, snakeweed, and tomatillo interspersed.

Monument No. 48. Chihuahuan Desert

MONUMENT NO. 49. Semidesert Grassland

Monument 49 was located in what was at that time an extensive tobosa flat. This biotic community, like that at Monument No. 27 and some others, could be classified either as a disjunct ecosystem within the extensive Chihuahuan Desert or as an isolated fragment of the semidesert grassland. As with the other similar habitat-induced "islands" of grass within a desert scrub matrix, I have considered this area as grassland.

The land on the Mexican side of the line here has been plowed and is now used to grow cotton. That on the U.S. side is tobosagrass, as the entire area was in 1892. There has been no apparent change in the vegetation west of the boundary since the earlier picture was taken ninety-two years ago.

MONUMENT NO. 49. Semidesert Grassland

Monument No. 50. Chihuahuan Desert

Although this area supports a rather typical stand of Chihuahuan Desert scrub today, with no grass, it was apparently characterized by a grass-scrub mixture in 1892. The ground cover between the bushes in the 1892 photo is grass, probably tobosa, or tobosa intermixed with sacaton. Even earlier than that the vegetation may have been composed largely or entirely of grasses, as the soil has a moderately dark cast indicative of grasslands and an extensive tobosa flat lies about a hundred yards distant.

The principal species growing in the area today are creosotebush, snakeweed, and tarbush with occasional individuals of banana yucca and prickly pear.

MONUMENT NO. 50. Chihuahuan Desert

Monument No. 51. Chihuahuan Desert

The earlier photo shows this as a typical tobosagrass swale that supported a dense stand of tobosa in 1892. The site retains its potential for growing tobosagrass today, but the grass has now largely disappeared and been replaced by snakeweed. The range on both sides of the line here is heavily grazed by cattle and horses with this resultant change in the vegetation. Although not shown in the recent photograph, mesquite is beginning to invade and, unless controlled, may be expected to continue to increase.

MONUMENT NO. 51. Chihuahuan Desert

Monument No. 52. Chihuahuan Desert

Insofar as can be determined from the original photograph, the vegetation here has changed little during the past ninety-two years. Although potentially a grassland site, it now supports a mixture of shrubs or half-shrubs, largely fourwing saltbush, wooly sagebrush, desert lavender, and desert hackberry. The sparse perennial grasses consist of galletagrass, threeawn, and black grama. An annual grass, six-week schismus, is also abundant.

The land on both sides of the line here is heavily grazed, with probable resultant depletion of the former palatable perennial grasses.

Monument No. 52. Chihuahuan Desert

Monument No. 53. Chihuahuan Desert

This is one of Emory's original monuments, and marks the southern end of the meridian portion of the boundary. These photographs were taken looking due north, consequently Mexico lies to the right, east of the monument, and the United States to the left. Monument 52 is barely visible in the repeat photo against the skyline immediately left of the tip of the monument.

The original picture shows a good stand of grass with scattered plants of sotol and occasional other small shrubs. Although the general aspect here in 1892 was grassland, it is now scrub.

Today the vegetation consists for the most part of honey mesquite, creosotebush, sotol, and snakeweed. Except for a single plant of sacaton, there are no perennial grasses in the vicinity although a nearby hill to the south does support a rather good stand of tobosagrass.

MONUMENT NO. 53. Chihuahuan Desert

Monument No. 54. Chihuahuan Desert

Starting the long trek due west along parallel 31°20′, the first monument is encountered on a shallow-soil upland site 2.47 miles from the meridian corner.

Except for stone monuments 40, 46, and 53, which were rebuilt the following year, monument erection along the meridian was completed in November 1892. It was not resumed on the parallel of latitude until June 16, 1893, and was completed September 19 of the same year.

The excessive drought that had gripped the Southwest during the preceding two years finally ended and, as given in Senate Document 247 Part II, 1898a, page 186:

> The monument party encountered this season an unprecedented rainy period. During the two preceding years hardly a drop of rain had fallen along this section of the boundary and in consequence the country had a burned barren appearance; nearly all the water courses had become dry; grass had disappeared; cattle by thousands had perished. This season the conditions were utterly changed; rain fell copiously and at very frequent intervals. The water courses were flooded, and the country took on a covering of green, in marked contrast to its former desert appearance. The few remaining cattle became fat and sleek, and grass grew in such quantities that vast fields were cut with machines and hay soon became cheap and abundant.

Although the vegetation in the vicinity of this monument is primarily scrub today, in 1892 it was grassland with scattered scrub species. Except for one plant of bush muhly beneath the fence, there were no grasses in the vicinity of the monument in 1984, although a tobosagrass swale can be seen about 200 yards beyond and to the left.

The former grass-shrub mixture has been replaced by annual forbs and a few bushes of tomatillo, tarbush, and Mormon tea. Most of the small shrubs now showing in the vicinity of the monument are tarbush.

Monument No. 54. Chihuahuan Desert

Monument No. 55. Chihuahuan Desert

This area would have been classified as grassland in 1892; today it has a shrub aspect with a thin stand of grasses scattered between the rocks and bushes. The monument, which is located on top of the ridge that shows in the earlier picture, is not visible. The former excellent stand of perennial grasses and scattered shrubs has now been replaced by a mixture of woody plants and a sparse stand of grasses. The dominant scrub species are little-leaf squawbush, sotol, honey mesquite, snakeweed, yerba de pasmo, desert lavender, and turpentine bush. The principal grasses are bullgrass, sideoats grama, and hairy grama, probably the same species in part that were dominant here originally.

Monument No. 55. Chihuahuan Desert

Monument Nos. 56–64. Semidesert Grassland

The area represented by these monuments, a distance along the boundary of about twenty-one miles, has changed little since the monuments were installed. The grasslands that prevailed at the end of the nineteenth century still do today, as do the scattered scrub species. It may be assumed that had there not been two years of severe drought immediately prior to the time the earlier photos were taken, the grasses would have been considerably more luxuriant then than they are today after ninety-two additional years of domestic livestock grazing.

With the exception of Monument 60, which is located in a sacaton flat at Antelope Wells, where sacaton is the only grass, gramas were the predominant identifiable grass species here when the repeat photo was taken in May 1984. Close grazing made identification of most of the grass plants difficult or impossible, and it can be assumed that several species in addition to those recorded were usually present. Note Monument No. 62, for example, where a second visit was made later in the year following the summer rains.

Keeping in mind the above comment on grass-identification difficulties, the following species were recorded as being most abundant at each of these nine monuments:

Monument No. 56. Grasses: hairy grama, black grama. *Woody species:* banana yucca, beargrass, prickly pear, sotol, honey mesquite, Mormon tea.

Monument No. 56. Semidesert Grassland

Monument No. 57. Grasses: blue grama, black grama. *Woody species:* honey mesquite, little-leaf squawbush, beargrass, sotol, yerba de pasmo, wooly sagebrush.

In the vicinity of this monument I encountered a large javelina (*Pecari tajacu* Linnaeus), with her young one, the first of the species encountered along the border to date. A short time later, in the same general area, we started up three mature antelope (*Antilocapra americana* Ord.), again the first we had encountered to date. Not long after we saw two more antelope and a doe mule deer (*Odocoileus hemionus* Rafinesque), once again the first we had seen.

MONUMENT NO. 57. Semidesert Grassland

Monument No. 58

Monument No. 58. Grasses: black grama, sideoats grama, threeawn, slim tridens. *Woody species:* little-leaf squawbush, sotol, beebrush, Mormon tea.

At this monument I saw another mule deer.

Monument No. 58. Semidesert Grassland

Monument No. 59

Monument No. 59. Grasses: black grama, sideoats grama, threeawn, slim tridens. *Woody species:* beebrush, honey mesquite. *Forbs:* croton.

Monument No. 59. Semidesert Grassland

Monument No. 60

Monument No. 60. Grasses: sacaton. *Woody species:* very occasional palmilla, honey mesquite. *Forbs:* Russian thistle, showing as the tangled mass between the camera station and the monument.

Monument No. 60. Semidesert Grassland

Monument No. 61

Monument No. 61. Grasses: threeawn, hairy grama, blue grama, plains bristlegrass. *Woody species:* honey mesquite, yerba de pasmo, threadleaf groundsel, Palmer agave.

MONUMENT NO. 61. Semidesert Grassland

Monument No. 62

Monument No. 62. Grasses: sideoats grama, hairy grama, black grama, galletagrass, tobosagrass, cane beardgrass, Texas beardgrass, plains lovegrass, wolftail. *Woody species:* sotol, western honey mesquite, buckwheat, wooly sagebrush, snakeweed. *Forbs:* croton.

MONUMENT NO. 62. Semidesert Grassland

Monument No. 63

Monument No. 63. Grasses: blue grama. *Woody species:* western honey mesquite, snakeweed.

Monument No. 63. Semidesert Grassland

Monument No. 64. Grasses: blue grama, sideoats grama, threeawn. *Woody species:* beargrass, western honey mesquite, wait-a-minute bush. *Forbs:* croton.

Monument 64 stands prominently in an excellent growth of beargrass that extends on both sides of the boundary. Although the grass cover here is good, the more prominent, large beargrass clumps dominate the landscape.

Monument No. 64. Semidesert Grassland

Monument No. 65. Evergreen Woodland

Monument 65 marks the highest point, 6,719 feet, along the entire boundary. It is located on a north-facing slope near the north end of the San Luis Mountains, a northern extension of the Sierra Madre of Mexico.

Although the trees that may have originally surrounded the monument site were probably cut down during construction or might have been destroyed by fire, the repeat photo seems to indicate a great increase in tree and brush cover during the intervening years. The background hills, which are in the United States and have received some fire protection for much of this time, show no indication of change.

The area surrounding the monument currently supports a dense stand of low-growing evergreen trees and chaparral. The dominant trees in the immediate vicinity are Mexican pinyon, alligator juniper, and silverleaf oak; the major chaparral species are mountain-mahogany, Mexican manzanita, Toumey oak, silktassel bush, and mountain yucca. The only grass species is an occasional plant of bullgrass.

The hills visible beyond the monument are covered by grasses growing between and beneath an open stand of trees or chaparral. The approximate mile of evergreen trees and chaparral in Mexico through which I walked to reach the monument contained remains of old charred stumps indicating at least one previous fire.

Monument No. 65. Evergreen Woodland

MONUMENT NOS. 66–69. Semidesert Grassland

These four monuments lie in the flat Animas Valley, spanning a distance of about nine miles between the San Luis Mountains on the east and the Guadalupe Mountains on the west. The Animas Valley was grassland when the monuments were established and remains so today.

Although the earlier pictures show an open grassland without trees or shrubs, there is no record of the kinds of grasses that prevailed. As I have indicated in the discussion of Monument 40, Mearns does not mention grasses by name, nor does he list or comment on the vegetation of Animas Valley.

Cooke (1938, p. 118), writing of his travels from 1846 to 1854, gives a poetic description of the valley looking toward the west and what was probably the Guadalupe Mountains, although he called them the Gilas:

> The mountains [the Animas or the San Luis] passed, before us was a smooth plain. . . . Waving with the south wind, the tall grama and buffalo grass received from the slant sunshine a golden sheen; and the whole had a rich blue and purple setting of long mountain ranges on either side.

Cooke was an army man, not a botanist, and throughout his journal grasses in New Mexico were either grama or buffalo. However, he leaves no doubt that the Animas Valley was well clothed with grasses of some sorts. Kearns (1906, p. 94) refers to it as "a grassy, treeless plain. . . ." Bartlett, writing of its appearance some forty years prior to erection of the monuments, commented: "So level was this valley, and so luxuriant the grass, that it resembled a vast meadow; yet its rich verdure seemed wasted, for no animals appeared, except a few antelopes and several dog-towns" (Bartlett, Vol. I, 1854).

As Cooke's party approached the Guadalupe Mountains,

> We passed very extensive prairie dog villages; in fact they lined the road all day; and I never remarked them before in apparently rich ground. The buffalo grass of late disputes predominance with the grama. The oaks, first descending from the mountains to the hills are now to be found even dotting the valleys, and we saw a very extensive grove to our left on the verge of the valley; there is also cherry. . . . Black-tailed deer and antelope are plenty; a number were killed (Cooke 1938, p. 119).

The Animas Valley today, although still basically grassland, has lost its pristine appearance and supports a mixture of grasses and annual forbs, the forbs being golden-eye and Russian thistle in large part. Much of it is heavily overgrazed by cattle, and though a few antelope may occasionally still be seen, they are outnumbered by cattle several hundred to one.

Monument No. 66. This monument is adjacent to an old reliable spring once called San Luis Springs but later and currently known as the Lang Ranch. According to Mearns (1906, pp. 92–93), this was a well-known camping place for the U.S. military in their wars against the Indians. He describes the location as it still is today:

> . . . close to the Boundary at Monument No. 66, just below timberline of the western foot of the San Luis Mountains, on the eastern edge of the broad Animas Valley. . . . Game was abundant, and water and grazing good. . . . game was so abundant that we killed seven antelope, two deer, two turkeys, two black timber wolves, and smaller game.

The two photographs of Monument 66 show the view looking northeast toward the San Luis Mountains. The 1892 photo shows a lush grassland extending to the base of the mountains where the grasses give way to oaks; the recent photo shows a less luxuriant grassland with a few cottonwoods at the old Lang Ranch site to the left of the monument. The oaks at the foot of the mountains appear much as they did ninety-two years earlier.

Only two grasses were identifiable here—blue grama and cane beardgrass. There may have been others, but a year of grazing since the last growing season made additional identification impossible. In addition to the grasses, a few patches of a forb, prickle poppy, had invaded the bare areas.

Monument No. 66. Semidesert Grassland

Monument No. 67

Monument No. 67. As at Monument 66, heavy grazing here made recognition of the grasses difficult. Despite this I was able to identify blue grama, a threeawn, sideoats grama, and cane beardgrass as the four probable principal species. A short distance north of the monument we startled two groups of antelope, three in one group and seven in the other.

Monument No. 67. Semidesert Grassland

Monument No. 68

Monument No. 68. The original photograph here shows no ground detail, but there is no reason to assume a ground cover of grasses appreciably different from that at the two preceding monuments. The repeat photo and my field notes show primarily grasses—squirreltail, blue grama, and cane beardgrass intermixed with the suffrutescent snakeweed and a forb, annual goldeneye, both of these latter two indicating deterioration of a once better stand of grasses.

Monument No. 68. Semidesert Grassland

Monument No. 69

Monument No. 69. Monument 69 lies on a slight rise near the western edge of Animas Valley. Three grass species predominate here—tobosagrass in the lower areas and squirreltail and galletagrass in the upland. A single low-growing shrub, wait-a-minute bush, occurs sparingly.

MONUMENT NO. 69. Semidesert Grassland

Monument No. 70. Evergreen Woodland, Grassland

This, the westernmost monument in New Mexico, lies high on a north-facing slope in the Guadalupe Mountains. From here one looks far to the west into Arizona. Because of the rugged terrain this monument could not be transported to the site entire and had to be carried there in sections on pack mules. The monument is still difficult to reach but would be even more so were it not for a jeep trail recently bulldozed into the hills by Mr. Cowan, the ranch owner. This enabled us to scramble in our VW bus to within about a half mile of the monument before continuing on foot.

The vegetation here seems to have changed little, if at all, since the monument was assembled and set up. Although low-growing trees and chaparral at first glance seem to dominate the area, there is also a moderately dense stand of perennial grasses. Two trees are common: one-seed juniper and black oak. Mountain-mahogany, beargrass, mountain yucca, and sotol are the principal chaparral species, plus the seemingly ubiquitous snakeweed. The grasses are well represented by sideoats grama, blue grama, black grama, bullgrass, and galletagrass.

Monument No. 70. Evergreen Woodland, Grassland

Monument No. 71. Evergreen Woodland

Located in the rugged Guadalupe Mountains, this monument was placed near the bottom of a steep slope where the line between Arizona and New Mexico intersects the International Boundary. The commissioners' report describes the difficulty of transporting the monument sections to the site but, typically, makes no mention of the vegetation.

A comparison of the two photographs shows no major vegetational changes except that some of the original trees are no longer there, others have come in, and some at least have grown. The oak to the left of the monument, for example, has grown considerably during the intervening years.

Two trees, Mexican blue oak and one-seed juniper, were common here when I visited the area in the summer of 1983. Numerically much more abundant, however, were many kinds of lower growing scrub species and perennial grasses. The most common scrub species were sotol, evergreen sumac, mountain yucca, turpentine bush, velvet mesquite, beargrass, feather dalea, and wooly sagebrush. The principal grasses were bullgrass, sideoats grama, wolftail, and galletagrass. The feather-like seedheads of bullgrass also show in the earlier picture.

MONUMENT NO. 71. Evergreen Woodland

Monument No. 72. Evergreen Woodland

Although only three-fourths of a surveyed mile west of No. 71, I reached No. 72 only after an hour of hot, hard climbing up and down the steep slopes. Here, too, the boundary commissioners in their report make no mention of the vegetation, saying only that the monument "was placed upon the summit of an extremely high, precipitous mountain" (Senate Document 247, Part II, p. 187).

Except for some replacement of individual plants by others there is no more discernible change in the general vegetation here than at Monument 71. Some of the oaks visible in the earlier picture are no longer there, and the foreground, which was formerly rather open with a few sotol plants, has been filled with a dense tangle of manzanita. Although the photographs suggest that there has been a major change from a tree-shrub mixture to shrubs alone, the general area still grows numerous trees, supporting the conclusion of little change.

Mexican manzanita and Emory oak are dominant here with huajillo, range ratany, amole, beargrass, and sotol interspersed between, and growing beneath, the trees. The same grasses occur here as at the preceding monument.

Monument No. 72. Evergreen Woodland

Monument No. 73. Riparian, Chihuahuan Desert

Monument 73 is located at the entrance to Guadalupe Canyon where the old nineteenth-century road began its tortuous climb up the canyon to the dreaded Guadalupe Pass. Two ecosystems are represented here, the riparian canyon bottom in the foreground and the more arid scrub-grass–covered hillside beyond.

A Mexican farm or ranch home is within hailing distance of the monument. This, and the usually plentiful water of the canyon stream, have resulted in long-continued and heavy grazing of the area in both Mexico and the United States.

Despite the grazing history, the riparian trees and shrubs have probably changed little during the intervening years and may be even more dense now than formerly. The background slope, on the other hand, appears to have evolved from an almost pure stand of grasses to a scrub-grass combination with scrub predominating.

The woody riparian vegetation consists for the most part of velvet mesquite, algerita, snakeweed, bird of paradise, desert broom, western hackberry, chittamwood, and threadleaf groundsel. Beneath and between the scrub species at the time of our visit was a good ground cover of ephemeral grasses and forbs. Arizona sycamore and velvet ash overhang the watercourse.

The scrub species visible on the hillside are largely ocotillo, little-leaf squawbush, velvet mesquite, beebrush, banana yucca, Palmer agave, and desert hackberry. Intermixed with these a few perennial grasses remain, mostly sideoats grama and tanglehead.

Monument No. 73. Riparian, Chihuahuan Desert

Monument Nos. 74–76. Chihuahuan Desert

Monument 74 is located on a hilltop two miles west of where the Guadalupe drainage heads south into Mexico at Monument 73. The next two (75 and 76) lie on a gently sloping plain or bajada east of the old San Bernardino Ranch.

Although none of these three shows any significant change in vegetation since they were erected, the amole at No. 74, although present in the earlier picture, has increased and now become dominant. A single plant of palmilla shows in the 1893 picture, but I recorded none near the monument in 1983. These, however, are minor, possibly individual plant differences, and suggest no change in life form.

Despite the moderately heavy use by cattle of the area in the vicinity of Monument 74, the rocky hilltop supports a variety of scrub species today. In addition to amole these are, for the most part, prickly pear, ocotillo, mariola, huajillo, feather dalea, spring whitethorn, snakeweed, desert zinnia, and twin-flower. The single species of perennial grass was grazed too closely to be recognizable.

The major species on the Mexican side of the line at Monument 75 in the summer of 1983 were snakeweed, velvet mesquite, and huajilla, with an occasional palmilla. These same scrub species also occurred across the fence in the United States. Here, however, grazing use had been much lighter, and two grasses were recorded—tobosagrass, which was widespread, and black grama, which occurred locally.

Although the general vegetational aspect had not changed at Monument 76, construction of the boundary fence has permitted marked differences in grazing pressure on the two sides of the line, resulting in fence-demarked plant species differences. On the more heavily grazed Mexican side, for example, the principal plants are snakeweed, creosotebush, tarbush, tomatillo, desert holly, spring whitethorn, prickly pear, and velvet mesquite. These same species also characterize the U.S. side, but to these are added gray coldenia and desert zinnia. Small amounts of closely grazed tobosagrass grow here south of the line; across the fence there is much more tobosa, intermixed with the highly palatable grass, bush muhly.

In the earlier picture a typical tarbush plant fills the left-lower quadrant. The other scrub species appear to be mostly tarbush and creosotebush.

Although Senate Document 247 (Part II, p. 187) briefly mentions the vegetation at Monuments 74, 75, and 76, the early photos belie the verbal description of *"a moderate growth of grass beside the usual varieties of cacti."* The pictures show no grass, although some surely was present. They also show no cacti, even though most of these, if they did occur, should have been readily distinguishable. No mention is made of the scrub-type vegetation, even though the photographs clearly show scrub as the dominant vegetation.

Monument No. 74. Chihuahuan Desert

MONUMENT NO. 75. Chihuahuan Desert

Monument No. 76. Chihuahuan Desert

Monument No. 77. Chihuahuan Desert

Senate Document 247 (Part II, p. 187) describes this boundary marker as located just west of the San Bernardino River. The 1958 U.S. Geological Survey College Peaks Quadrangle map shows no San Bernardino River but names the first drainage east of the monument at Black Draw. The two, I assume, are identical. Although the document does not mention the vegetation surrounding the monument, it does say "The San Bernardino Valley at this place is flat and marshy, covered with grass, and has several springs of good water, also a number of quite large pools." This description agrees with those of others who had passed through here earlier.

The monument itself is located on a dry site a few feet above the drainage a short distance west of the swampy area referred to above. At the time of the earlier photo it was surrounded, as it still is, by a dense stand of brush. The same species probably predominated then that do today. When I saw the area in the summer of 1983 these consisted of an almost pure overstory of velvet mesquite with a scattering of four-wing saltbush and an understory of snakeweed. There were no perennial grasses.

Monument No. 77. Chihuahuan Desert

Monument Nos. 78 and 79. Chihuahuan Desert

These two monuments are located on a gently sloping bajada that rises gradually to the Perilla Mountains on the west. Both lie in typical upland Chihuahuan Desert vegetation that seems to have changed little or not at all during the ninety-two years that have elapsed since the monuments were erected.

Except for ephemerals that come and go with seasons and years, the vegetation here consists exclusively of scrub species. The most abundant of these are creosotebush, tarbush, spring whitethorn, mariola, Englemann prickly pear, snakeweed, desert holly, and gray coldenia. I recorded no perennial grasses at either monument, nor are any distinguishable in the earlier photograph.

MONUMENT NO. 78. Chihuahuan Desert

MONUMENT NO. 79. Chihuahuan Desert

MONUMENT NO. 80. Chihuahuan Desert

Located on a high, north-facing ridge of the Perilla Mountains, Monument 80 provides distant views both to the east and to the west.

A comparison of the two photographs shows almost complete annihilation of a once good stand of grass here, accompanied by an increase of scrub vegetation. Although a little ocotillo and a few other shrubs show in the earlier picture, a good blanket of perennial grasses formerly characterized most of the visible area.

A few grass plants and many shrubs grow today on the background hill and in the immediate vicinity of the monument; however, shrubs or halfshrubs now greatly predominate. Two species, whitethorn acacia and snakeweed, are most abundant, but are intermixed with prickly pear, cane cholla, little-leaf squawbush, Palmer agave, velvet mesquite, banana yucca, odora, and huajilla. The only two grasses that I recorded in 1983 were very minor amounts of curly mesquite and threeawn.

MONUMENT NO. 80. Chihuahuan Desert

MONUMENT NO. 81. Chihuahuan Desert

Located on the highest point where the boundary crosses the Perilla Mountains, and just .83 of a mile west of Monument 80, No. 81 was reached only after thirty-five minutes of hiking over steep terrain. An old, washed-out trail, once passable for four-wheel-drive vehicles, would not have been usable even for these in 1983.

The 1893 photo indicates only scrub vegetation but shows too little area and detail to permit drawing any conclusions as to the presence or absence of grasses. However, the proximity of this monument to No. 80 and the similarity of habitat, land use, and climate lead me to conclude that here, too, there has been a marked decrease in grasses and a concurrent increase in shrubs. During the intervening years the woody species have apparently increased in number, as they certainly have in size. Scrub species predominate here today, principally whitethorn acacia and Englemann prickly pear. The large bushes that show on both sides of the monument are whitethorn acacia. Additional scrub species include ocotillo, banana yucca, and Palmer agave. These are accompanied by two forbs, one of these a species of mallow (*Sphaeralcea*), the other a perennial buckwheat.

There were more grasses here than at Monument 80, particularly on the U.S. side of the boundary fence. On either side, however, I recorded only two grass species: sideoats grama and bush muhly.

Monument No. 81. Chihuahuan Desert

Monument No. 82. Chihuahuan Desert

About 1.1 mile west of Monument 81, Boundary Marker 82 lies on the western slope of the Perilla Mountains, overlooking the smelter town of Douglas and the Sulphur Springs Valley. Shown in the background of both pictures is a prominent landmark, Cerro Gallardo, or "Niggerhead" as it was known to the monument crew. This monument was built on the site of Emory's old Monument 19, which had been constructed ca. 1855.

The ninety years that have elapsed since 1893 have shown a shift from grasses with a very few scattered shrubs to shrubs and halfshrubs with no grasses. One of the two recognizable woody species in the 1893 photo, sotol, occurs nowhere in the vicinity today. Ocotillo, which formerly occurred sparsely, has now become much more abundant. In addition to the ocotillo, other shrubs common here today are snakeweed, huajilla, little-leaf squawbush, Palmer agave, and Wright beebrush. These are interspersed with lesser amounts of range ratany, crucillo, spring whitethorn, wait-a-minute bush, and Mormon tea. This entire area, both north and south of the boundary fence, appears to have been heavily grazed for many years and, in range parlance, is in "poor condition."

MONUMENT NO. 82. Chihuahuan Desert

Monument No. 83. Chihuahuan Desert

A Border Patrol road of sorts extends from Douglas on the west to a point a few miles east of Monument 83, enabling us to drive to within a few feet of this monument. The monument was placed on the north side of a small, rocky hill which, despite its rocky character, is heavily grazed on both sides of the line.

If there has been any major change in the vegetation since 1893, the pictures do not show it, the dominant plants then, as now, consisting of scrub species. As at Monument 82, sotol, which was conspicuous in 1893, has vanished, and does not now occur in the immediate vicinity of the marker. Today, the vegetation consists for the most part of velvet mesquite, little-leaf squawbush, desert hackberry, crucillo, Wright beebrush, and snakeweed, with lesser amounts of Mormon tea, Wright desert holly, and trixis. The only perennial grass was an occasional bush muhly growing beneath the bushes or in the rocks where it was protected from the ever-hungry cattle.

Monument No. 83. Chihuahuan Desert

MONUMENT NO. 84. Chihuahuan Desert

At the time the monuments were erected, the towns of Douglas in the United States, and Agua Prieta across the line in Mexico, did not yet exist. As a consequence, the vegetation in the vicinity of Monument 84, which is adjacent to the Douglas airport, has changed greatly during the intervening years. At first glance, because of the dissimilarities in the background and vegetation, the two photographs may appear to represent two different areas. They are, however, views of the same monument, both taken facing to the east. In the earlier picture Cerro Gallardo may be seen in its entirety to the right of the monument; in the repeat photo the top of the mountain is hidden by smoke from the copper smelter in Douglas.

The 1893 photo shows this area with rather typical Chihuahuan Desert scrub vegetation with creosotebush and tarbush appearing to predominate. Two banana yucca plants can be seen to the right of the monument. This vegetation has since been cleared on the Mexican side of the line and, obviously, where the road has been graded across the fence. North of the road, however, creosotebush, tarbush, mariola, and desert zinnia now predominate. The shrubs that show south of, and adjacent to, the fence are creosotebushes that are becoming reestablished. Between the widely scattered bushes is a ground cover of ephemerals, principally members of the grass, mustard, and plantain families, intermixed with Russian thistle.

Monument No. 84. Chihuahuan Desert

Monument No. 85. Urban

What was formerly a broad, unobstructed expanse of the Sulphur Springs Valley is now a city street in Agua Prieta. Although not visible through the fence, an equally radical change has occurred in Douglas, across the International Line.

The early photo shows an open stand of Chihuahuan Desert scrub that probably consisted mostly of creosotebush, tarbush, and mariola. Between the island-like stands of shrubs was an open matrix of forbs or grasses. As the picture was taken in the summer during a year of unusually heavy rains, the ground cover may also have contained many ephemerals.

In contrast, the little vegetation able to grow here today consists of a few individuals of four-wing saltbush adjacent to the fence and the always opportunistic Russian thistle on a little bare soil between the street and the sidewalk.

MONUMENT NO. 85. Urban

MONUMENT NOS. 86–88. Chihuahuan Desert

Three scrub species occur commonly at Monuments 86, 87, and 88—velvet mesquite, spring whitethorn, and tarbush. At Monuments 86 and 87 these three occur to the almost total exclusion of all other vegetation; at Monument 88, as noted below, several other shrubs and halfshrubs are also abundant.

Although the shrubs today are larger and apparently growing more luxuriantly than in 1893, there is no evidence of a change in either life form or composition at any of these three monuments during the intervening years. A partial ground cover of grasses or forbs can be distinguished in the earlier pictures of Monuments 86 and 87 where there is none today. Vigorous plants of spring whitethorn are shown growing on the fence line at these two monuments, in one instance partially obscuring Monument 86.

As the 1893 photos were taken during the summer rainy period of a year with exceptionally heavy precipitation, the ground cover probably consisted, at least in part, of ephemeral forbs. And, because perennial grasses were growing at many of the other nearby monuments where the habitat was similar, it is probable that at least some of the ground cover consisted of grasses. Although I recorded no grasses at either of these locations in the summer of 1983, tobosagrass did occur in nearby swales and drainages.

The vegetation at Monument 88 consists currently of typical Chihuahuan Desert scrub species, but is floristically considerably richer than at the two preceding markers. In addition to the three shrubs that characterized Monuments 86 and 87, creosotebush, mariola, Wright beebrush, little-leaf squawbush, huajilla, sotol, desert zinnia, shrubby senna, and menodora are also common here.

Monument No. 86. Chihuahuan Desert

MONUMENT NO. 87. Chihuahuan Desert

MONUMENT NO. 88. Chihuahuan Desert

Monument No. 89. Chihuahuan Desert, Semidesert Grassland

Monument 89 is located on the crest of a small rocky hill where the vegetation is a mixture of perennial grasses and scrub species in a vegetation type that is often classified as upper desert grassland. Analysis of the 1893 and 1983 pictures reveals no apparent change in life form here during the time interval represented.

Despite a long-time history of cattle raising throughout this area, very few perennial grass plants have survived on either side of the boundary fence, and I recorded only occasional individuals of two grass species in the summer of 1984: slender grama and tanglehead. As sites of this sort and with similar precipitation are capable of growing numerous other perennial grasses, these two probably represent only relicts of a once much richer grass flora.

Several shrubs or halfshrubs are now much more conspicuous at this monument than the grasses, namely huajilla, ocotillo (also distinguishable in the earlier photo), Palmer agave, southwestern twinflower, croton, and wooly sagebrush.

MONUMENT NO. 89. Chihuahuan Desert, Semidesert Grassland

Monument No. 90. Chihuahuan Desert

A graded highway north of the boundary parallels the International Line here, destroying the vegetation. Adjacent to the highway however, as well as across the fence in Mexico, the native woody cover remains. Although there are no grasses growing here today, there is a moderately dense stand of scrub, comprising primarily creosotebush, spring whitethorn, tarbush, and gray coldenia. These are intermixed with lesser amounts of little-leaf squawbush and desert zinnia.

There is some low-growing vegetation suggesting perennial grasses visible in the earlier picture, as well as several yucca-type plants. In contrast, I noted neither grasses nor yuccas when I took the repeat photo. Even though there may previously have been some perennial grasses, a scrub type of vegetation dominated the area in 1893 as it still does today. There has, therefore, been no major change in life form, even though the scrub species do appear to be considerably more abundant and larger now than formerly.

Monument No. 90. Chihuahuan Desert

Monument No. 91. Chihuahuan Desert

Monument 91 stands on the crest of a narrow limestone ridge about a mile south of the small town of Bisbee Junction.

Little vegetation grew on the rocky ridge in 1893; little more grows there today. Near the base of the bluff, however, and in the surrounding area, there is an abundance of typical Chihuahuan Desert scrub species. Immediately adjacent to the monument are scrub oak, mariola, little-leaf squawbush, and Wright beebrush. The rocks make access difficult for domestic livestock and one grass species, sideoats grama, may be found between the shrubs. Near the base of the ridge all of the above scrub species are common and, in addition, crucillo, beargrass, feather dalea, croton, and range ratany.

The early photograph shows too little of the area and its vegetation to permit drawing any conclusions regarding possible changes in either life form or composition of the plant cover.

MONUMENT NO. 91. Chihuahuan Desert

Monument No. 92. Semiurban (L), Reseeded Grassland (R)

When the monuments were erected, this was open, uninhabited terrain. Today the town of Naco lies a half mile west of the monument, and man's influence is evident both north and south of the International Line. What once appeared as a mesquite-grassland mixture is now largely a weed patch of Russian thistle and Thurber mustard south of the line, and an artificially seeded, almost pure Lehmann lovegrass pasture on the north. The brush along the fenceline is mostly velvet mesquite.

Monument No. 92. Semiurban (L), Reseeded Grassland (R)

Monument No. 93. Semidesert Grassland, Chihuahuan Desert

This view, taken facing to the east, shows Naco in the distant background. A ranch road parallels the fence here on the Mexican side of the line and a Border Patrol/ranch road on the United States side. Construction and use of these roads have partially or completely destroyed the vegetation in their rights-of-way. Elsewhere in the vicinity of the monument the vegetation can be classified as "natural."

There has been little apparent change in life form here during the past ninety years. Scrub species characterized the foreground in 1893, as they do today. The distant background in the earlier picture appears to be an extensive grassland; today the grasses south of the line have been replaced largely by ephemeral forbs. The two-to-three-foot-high shrubs in the recent picture are either honey mesquite or a low-growing form of velvet mesquite. Between these, and showing in the foreground as dark-color, ground-hugging vegetation is a second, very common woody species, huajilla. Both of these shrubs are abundant today, both north and south of the fenceline.

In addition to the shrubs, two grasses—Santa Rita threeawn and bush muhly—and two forbs—desert holly and large-flowered zinnia—occur here sparsely today.

MONUMENT NO. 93. Semidesert Grassland, Chihuahuan Desert

Monument Nos. 94 and 95. Semidesert Grassland

The extensive plain where these monuments were erected was, and still is, largely grassland, dotted with an occasional bush of velvet mesquite. Except for some almost certain differences in species, the vegetation here seems to have changed little since the earlier pictures were taken.

The low-growing huajilla, which persists even under heavy grazing, is abundant throughout this portion of the boundary. It appears to constitute the principal ground cover in the 1893 photos, as it still does. A third shrub, velvet-pod mimosa, also occurs here, but less frequently than either the mesquite or the huajilla.

Although the shrubs tend to be conspicuous, most of the vegetation here today consists of grasses, principally red threeawn, Lehmann lovegrass, plains bristlegrass, Arizona cottongrass, and blue grama. All of these with the exception of Lehmann lovegrass, a comparatively recent introduction from South Africa, were almost certainly present in 1893 as members of an even more extensive grass flora. This grass dominance is suggested by the soil, a red clay, that holds moisture well and which, as on the Santa Rita Experimental Range between Nogales and Tucson, Arizona, characteristically supports a good cover of perennial grasses.

Monument No. 94. Semidesert Grassland

Monument No. 95. Semidesert Grassland

Monument Nos. 96 and 97. Semidesert Grassland, Scrub

The vegetation continues here as part of the same ecosystem that prevails eastward almost to Monument 91. The principal species differ somewhat from marker to marker, but the dominant life forms and overall habitat remain the same. The Huachuca Mountains show in the background; the San Pedro River, which lies between the monuments and the mountains, is not visible.

The shrubs that show beyond the monuments in the 1893 photographs appear to be mesquite for the most part, as they are today. Other scrub species of about the same height—principally spring whitethorn, littleleaf squawbush, crucillo, and Mormon tea—occur in the area today and may also have been there when the earlier pictures were taken. Although not visible in any of the pictures, banana yucca, palmilla, and range ratany also grow in the vicinity today and were probably also there at the time the monuments were erected.

In the repeat photo of Monument 96 the mesquite and other scrub species seem to be about as abundant south of the line as they were formerly; on the north they occur relatively sparsely. This sparseness is due, not to natural factors, but to recent "range improvement" practices that have largely eradicated the various woody plants.

The earlier pictures show what appear to be closely grazed grasses in the foreground. Although not visible in the photographs, these would also have been growing between the bushes throughout most of this area. The pictures were taken in late June or early July 1893. The July, August, and September rains that year were exceptionally heavy, and had the photographs been taken in late summer, after the grasses had matured, the entire aspect would have been markedly different. The repeat photos were taken in mid-June 1983, again almost a year after the previous year's growing season. Thus, they too, after almost a year of grazing, show few grasses but an abundance of the comparatively unpalatable forbs.

I recorded only two perennial grasses here: tobosa and red threeawn, neither of which is particularly relished by cattle and consequently might be expected to survive longer under grazing pressure than other, more palatable species. Intermixed with the grasses, and occurring much more abundantly, were two ephemeral forbs—Russian thistle and Thurber mustard. A perennial forb, desert holly, was common locally, particularly beneath the shrubs.

Except for the recent eradication of woody plants north of the boundary here, none of the available evidence suggests that there have been any major vegetational changes during the past ninety years. There have undoubtedly been changes in composition, particularly of the grasses, but these cannot be determined from the pictures.

Monument No. 96. Semidesert Grassland, Scrub

MONUMENT NO. 97. Semidesert Grassland, Scrub

SAN PEDRO RIVER

The San Pedro River, the easternmost of south-central Arizona's two so-called rivers, both of which flow mainly to the north, crosses the boundary .35 miles east of Monument 98.

The San Pedro River at the International Boundary is described in Senate Document 247, page 188 (1898a):

> The bed of the stream has been sunk by the attrition of the current to 8 to 15 feet below the surface of the ground, and is from 30 to 60 feet in width. In ordinary seasons but little water is found in the stream, but during the operations of the monument party in this vicinity, heavy floods caused the river frequently to rise bank full, and as there are no bridges its depth at times seriously interrupted communications between opposite banks.

Mearns (1907) describes the river at Monument 98 (p. 101) as:

> a good-sized stream, containing many batrachians and turtles, and its waters and banks are inhabited by numerous species of mammals and birds. There are, however, no meadows or marshes of any considerable extent along its banks in this part of its course. Trees are limited to the edge of the stream, where willow, ash, boxelder, cottonwood, and mesquite are the common species.

Cooke (1938), in his account of the San Pedro at about this point in 1846–54, comments on the fish in the river: "Fish are abundant in this pretty stream. Salmon trout are caught by the men in great number; I have seen them 18 inches long." These, according to Dr. Donald A. Thomson, curator of fishes, University of Arizona, were Colorado River squawfish (*Ptychocheilus lucius*). As Cooke's party began to travel north down the San Pedro from this point, Cooke reported "The road today was quite crooked and rather difficult to open, the bottoms having very high grass and being lumpy." This, quite obviously, was sacaton, a tall grass that would have provided an extremely "lumpy" road for the wagons to travel.

A description today of the San Pedro River where it crosses the boundary would not differ greatly from those of the early travelers. The river at this point had cut a definite channel at least as early as 1850 and probably has changed little since then. Immediately north of the boundary fence I found a ten-to-twelve-foot vertical drop on the east bank in 1983 and 1984; on the west bank a drop of two to three feet. Elsewhere in the area the depth of the channel varied but was generally less than at the fence line. There were few signs of recent erosion, suggesting on the average, a long-stabilized, non-eroding streambed.

When I visited the area in June 1983, and again in September 1984, the river was not in flood (nor were there any large fish); instead a quiet, clear stream a few inches deep and up to about five feet across provided a placid habitat for large numbers of small minnows, usually no more than an inch in length. The banks were lined with large Fremont cottonwoods and black willows, with many thickets of arrow weed and tamarisk. Sacaton, with its "lumps" of tall grass, still covered much of the floodplain area above the streambed.

Monument No. 98. Semidesert Grassland

Monument 98, located a short distance west of the San Pedro River, was rebuilt from the material used to construct Emory's old Monument No. 20. It stands on the north bank of a normally dry tributary to the river. Thus, the drainage that shows in the pictures is the tributary, not the river.

There has been a marked increase in the woody vegetation of the drainage since the first picture was taken. Grasses, which still characterize the upland interdrainage areas here, formerly extended across the waterways as well, but were probably represented in part by different species. Velvet mesquite makes up most of the woody vegetation in the drainage today, although with an occasional palmilla and Mormon tea.

The interdrainage upland north of the boundary, from the San Pedro River to the foothills of the Huachuca Mountains, has recently been seeded to Boer lovegrass. This has proved to be well adapted, has become the principal grass over most of this area, and is now spreading into Mexico. Blue grama is also abundant but is being hard pressed by the relatively aggressive lovegrass. Nongrass species that are common here include range ratany, shrubby buckwheat, croton, sotol, and velvet mesquite. Most of the interdrainage mesquites are small, indicating recent encroachment into the grassland. A young mesquite shows in the left foreground of the repeat photo.

Monument No. 98. Semidesert Grassland

Monument No. 99. Semidesert Grassland

These pictures, which were taken facing east toward the San Pedro River, show an essentially unbroken grassland in the earlier photo; a grassland being invaded by scrub in the repeat picture. The rangeland on both sides of the boundary fence here is being taken over by woody species, but primarily by velvet mesquite. A typical mesquite shows to the right of the monument in the repeat photo.

As at Monument 98, the principal grasses here are the exotic Boer lovegrass and the endemic blue grama. These are interspersed with an abundance of huajilla. This shows as the dark-colored, low-growing plants in the repeat photo; several may be seen immediately west of the monument.

Monument No. 99. Semidesert Grassland

Monument Nos. 100 and 101. Semidesert Grassland, Evergreen Woodland

These boundary markers are located on steeply sloping south-facing hillsides near the southern end of the Huachuca Mountains. Although no change in life form is evident since the earlier pictures were taken, the vegetation today appears much more luxuriant than in the earlier photos. The relative dearth of vegetation in 1893 may have been due to the preceding years of drought or to a recent fire. Whatever the reason, there was a poor ground cover at the time Monument 100 was erected. By comparison, the repeat photo of Monument 100 was taken in September of 1984, following an exceptionally rainy summer, and shows an excellent stand of grasses.

I encountered no richer grass flora along the entire boundary than I did here. Ten grass species in all were recorded: tanglehead, plains lovegrass, cane beardgrass, squirreltail, bullgrass, plains bristlegrass, Arizona cottongrass, blue grama, sideoats grama, and a species of the genus *Hilaria* that I identified (perhaps mistakenly) as galletagrass.

Interspersed among the grasses were a number of woody plants, primarily Wright beebrush, fern acacia, kidneywood, sotol, Palmer agave, and squawbush. The dark-colored shrub to the right of the monument is squawbush. The slopes above both monuments, as well as the draws in the area, support scattered, vigorous trees of Mexican blue oak. Some of these show against the skyline in the photo of Monument 100. The same photo shows an unidentified composite forb.

Despite the disparity in appearance of the vegetation in the earlier and later pictures, there has been no apparent life form change here. The boundary in this area lies at the lower limits of the evergreen woodland and many of the grasses noted are characteristic of that vegetation type. This was apparently the situation in 1893, as it still is today.

MONUMENT NO. 100. Semidesert Grassland, Evergreen Woodland

Monument No. 101. Semidesert Grassland, Evergreen Woodland

Monument No. 102. Semidesert Grassland, Evergreen Woodland

Located on the western slope of the Huachuca Mountains, this boundary marker provides an overview across the extensive Santa Cruz River Valley as far as the Patagonia Mountains. Little grass shows in the original picture, probably reflecting the severe drought of the previous two years.

The photographs indicate a strong shift in the foreground from vegetation consisting largely of sotol to a mixture of grasses, prickly pear cactus, and ocotillo. Although this shift is true to some extent of the general area, sotol is still abundant nearby. There has, however, been a marked increase nearby in ocotillos and prickly pear. Grasses, also, are as prevalent as the picture indicates. The prickly pear and ocotillo suggest either deterioration of the area as a forage-producing site or a shift to species with greater drought tolerance. The grasses, conversely, suggest an increase in rainfall. These apparent conflicts, combined with the evidence from Monuments 100 and 101, lead me to conclude that the comparative paucity of the grasses in 1893 was a result of the previous drought and/or fire. Permanent water for livestock is currently available less than a mile away, making the vegetation here subject to close grazing.

In addition to the three species mentioned above, other shrubs, principally huajilla, fern acacia, cane cholla, squawbush, mountain-mahogany, and banana yucca are common in the vicinity of this monument. Two species of oak, Emory and Mexican blue, are dominant nearby. These, together with squawbush, constituted the dark vegetation on the skyline of the original picture. These same species are locally visible on the skyline of the repeat picture.

The grass flora at this site was considerably less varied when I visited this monument in 1983 and 1984 than it was at the two previous monuments. Three species—sideoats grama, blue grama, and cane beardgrass—were the most common.

MONUMENT NO. 102. Semidesert Grassland, Evergreen Woodland

Monument Nos. 103–5. Evergreen Woodland, Semidesert Grassland

These, the first three monuments west of the Huachuca Mountains, lie in an open woodland that adjoins open semidesert grassland along its lower altitudinal limits. Two species of oak characterize the tree cover: Emory and Mexican blue. Velvet mesquite and alligator juniper occur only occasionally. Perennial grasses constitute most of the understory.

Although this savanna rangeland has been subjected to long-continued, and often heavy, grazing, the numbers of livestock have been controlled for many years, and a rather wide variety of perennial grasses currently grows here. I recorded ten species in all at one or more of these three monuments in 1983 and 1984. These consisted of plains lovegrass, sideoats grama, hairy grama, spruce-top grama, purple grama, cane beardgrass, wolfspike, curly mesquite, wolftail, and vine panic.

The pictures give no indication of any life form change here during the intervening years. Individual trees have gone (and come) and, at Monument 103, the stand of oaks may now be denser than before. Grass growth has also improved greatly but this may be attributed to the two years of drought that preceded 1893 and recent heavy summer rains that fell before the 1984 picture was taken.

MONUMENT NO. 103. Evergreen Woodland, Semidesert Grassland

MONUMENT NO. 104. Evergreen Woodland, Semidesert Grassland

Monument No. 105. Evergreen Woodland, Semidesert Grassland

Monument Nos. 106 and 107. Semidesert Grassland

Boundary Marker 106, constructed at the site of original Monument No. 21, stands prominently on a grass-covered expanse of Campini Mesa. Senate Document 247 (1898a) places the monument about a kilometer west of the line dividing Pima and Cochise counties. This is an apparent error, as the U.S.G.S. Sunnyside Quadrangle map (1958) shows it about 150 meters east of this line. The photographs of both monuments were taken facing west and show the Patagonia Mountains in the background.

There has been no apparent change in life form here since the monuments were erected in 1893. Perennial grasses constituted the dominant life form then, as they still do. Careful study of the original pictures reveals a moderately dense ground cover of grasses, despite the severe drought of the two previous years. Extremely close grazing by the large numbers of cattle that were in the area in the 1890s, and the poor quality of the original photograph of Monument 107, tend to give the false impression of little or no vegetation at that time.

Four species of perennial grasses dominate the area today: blue grama, sideoats grama, plains lovegrass, and curly mesquite. A perennial half-shrub, threadleaf groundsel, is moderately abundant, particularly north of the line. A few oaks, both Mexican blue and Emory, show in the background of both monuments today as well as in the earlier pictures.

Monument No. 106. Semidesert Grassland

Monument No. 107. Semidesert Grassland

Monument No. 108. Semidesert Grassland, Evergreen Woodland

This masonry monument was rebuilt at the site of old Monument No. 22. It stands on the west bank of Bodie Canyon, a seasonally dry, major tributary of the Santa Cruz River.

This area remains today, as it was in 1893, a savanna type with oaks as the tree component of the vegetation. There has been no change in life form with the passage of time, although some of the trees have matured. The small individuals to the immediate right of the monument in the early photo have now become full-grown trees. Controlled grazing, combined with fire control in recent years, have benefitted both the grasses and the trees.

Although two species of oaks, Emory and Mexican blue, are common along the washes of this general area, only Emory oak shows in the recent photograph. The understory grasses are principally sideoats grama, blue grama, hairy grama, sprucetop grama, cane beardgrass, wolfspike, bullgrass, and curly mesquite. Not showing in the picture but common in the adjacent riparian habitat of the wash are Arizona sycamore and velvet ash.

MONUMENT NO. 108. Semidesert Grassland, Evergreen Woodland

Monument Nos. 109 and 110. Semidesert Grassland

These two monuments are located in the open, treeless expanse of the San Rafael Valley. The Patagonia Mountains show in the background of Monument 109; Parker Canyon drainage lies immediately east of Monument 110.

Grasses remain the characteristic life form here today as they were when the monuments were erected. Occasional oaks dotted the drainages then and still do today. The perennial grass flora is much more restricted than at Monument 108, consisting largely of blue grama and an occasional sideoats grama. There may have been some other shortgrasses at these monuments when I visited them in 1983, but removal of the seed heads by close grazing made further identification impossible.

The dark-colored clumps of vegetation that can be seen on the slope east of Monument 110 are primarily Emory oak. Scattered young plants of velvet mesquite are becoming established in the area although none show in the photographs.

Monument No. 109. Semidesert Grassland

Monument No. 110. Semidesert Grassland

Monument No. 111. Semidesert Grassland

Monument 111, built from the remains of the original Emory Monument No. 23, stands on a rise of ground just west of the Santa Cruz River. The Santa Cruz, which originates in Arizona, first flows south around the Patagonia Mountains, then turns north and reenters Arizona a few miles east of Nogales. Thus, the river intersects the International Boundary at two locations.

The photographs were taken facing northeast and show the Huachuca Mountains in the background. The repeat picture was taken from a camera location closer to the monuments than in the original and tends to look down on the river-bottom trees that do not show in the 1893 photo. Early accounts indicate that trees grew abundantly along the banks of the river, and whether those showing in my repeat picture were there, but were much smaller in 1893, is a moot question. If they were, they have grown considerably taller during the intervening years. They may also show more in the recent picture because of the different camera angle.

The original picture shows no grasses, a condition that almost certainly reflects the previous severe drought and heavy overgrazing. Today, the area in the vicinity of the monument, as generally throughout this valley area, is vegetated with a moderate to good cover of perennial grasses, mostly blue and sideoats gramas. The adobe walls showing in the original picture have returned to their native clay, and the residual mound that marks their former location now supports a stand of sacaton.

Senate Document 247 makes no comment on the Santa Cruz River at this location. Today, however, there is a small, clear stream in a shallow channel that is not characterized by the deep cutting that typified the San Pedro at the boundary. The floodplain here supports a good stand of sacaton as it must have at the time the monuments were erected. Flooding has cut into this on the west bank immediately north of the boundary fence, otherwise there has been little erosion. Large Fremont cottonwoods and black willows border the stream, particularly on the eastern bank.

MONUMENT NO. 111. Semidesert Grassland

Monument No. 112. Semidesert Grassland

This area, which I have designated *semidesert grassland,* is rapidly being taken over by velvet mesquite and might, perhaps, be more properly classified as savanna. An occasional Emory oak lends variety to the mesquite. The oaks are probably about as abundant as they were in the 1890s; the mesquite, in contrast, presents a marked life-form change. What was essentially open grassland has now become a mesquite-grass savanna in which the mesquites are assuming an ever-increasing greater control. Their root systems, which are both deep and extensive, deprive the grasses of needed moisture in a land where the limited life-sustaining moisture is most available to the most aggressive and best equipped life forms or species.

Two grasses predominate here today: blue grama and sideoats grama. Several forbs are also common, principally spiny burroweed, ragweed, large-flowered zinnia, and an annual plantain.

Monument No. 112. Semidesert Grassland

Monument Nos. 113 and 114. Evergreen Woodland

These two boundary markers are in the Patagonia Mountains, about three miles south of the picturesque, abandoned mining settlement of Duquesne. Both monuments, but No. 114 in particular, are located in rough, mountainous terrain and are not accessible by road.

The original pictures of both monuments show the ground so nearly devoid of vegetation as to suggest the occurrence of a recent fire. Whether as a result of fire, or of drought combined with heavy grazing, there was little ground cover when the monuments were erected. There was an overstory of Emory oaks, as there is today. Today, however, although there has been no change in life form, the oaks are more abundant than before, particularly at Monument 113, and have grown considerably larger.

Most of the grasses at these monuments are rather tall, coarse-growing species that typically characterize this oak-savanna vegetation type in south-central Arizona. The principal grass species when I visited the area in July 1983 were mountain muhly, bull grass, deergrass, and cane beardgrass. It may be assumed that these same grasses characterized the area in 1893. Other common plants in the area today are the large-leaved, sprawling forb ipomoea, shrubby buckwheat, and (at No. 114) sotol and kidneywood.

Monument No. 113. Evergreen Woodland

Monument No. 114. Evergreen Woodland

MONUMENT NO. 115. Semidesert Grassland

Monument No. 115 proved to be difficult to reach, requiring several miles of hiking. This terminated in a thousand-foot (vertically) climb up a rock-strewn, well-grassed mountain lying just west of the main Patagonia Mountain mass.

The 1893 picture, like those of the four preceding monuments, seems to show the effects of a recent fire. As high as this monument site is, and its considerable distance from and elevation above any possible source of water for livestock, makes it improbable that overgrazing alone could have given the area such a denuded appearance. Even today, this isolated mountain may be burned periodically, as suggested by the almost total lack of trees or shrubs at an altitude where woody species might be expected to predominate. Burned sotol stumps in February 1984 provided evidence that at least one fire had swept through the area within the last few years.

The remains of a few shrubs or low-stature trees are visible on the hill to the left of the monument in the earlier picture. These are leafless, even though photographed in June, and appear to have been killed by fire. The same slope today supports only grasses. Despite these few woody plants, however, there is no evidence of any major change in life form, and it may be assumed that this area was formerly characterized largely by grasses, as it is entirely today. Emory oak, velvet mesquite, and ocotillo occur in the drainages at the base of the mountain where fuel is insufficient to carry fires, but none of these species extends for any distance up the slopes.

The high elevation of this land mass and its steeply sloping sides serve to protect the grasses here from much grazing by domestic livestock. This, and its apparent susceptibility to burning, favor the grasses, with the result that the vegetation here represents what may be termed the *upper desert grassland* at its optimum.

The principal grasses in the vicinity of the monument today are sideoats grama, plains lovegrass, Texas beardgrass, cane beardgrass, bullgrass, threeawn, and curly mesquite. Scattered between the dominant grasses are sotol and huajilla and wait-a-minute bush. These last two are so-called *sprouters* after fire and probably are as resistant to fire as are the grasses.

Monument No. 115. Semidesert Grassland

Monument Nos. 116 and 117. Semidesert Grassland

These two monuments were erected on the lands that slope westerly from the Patagonia Mountains to the Santa Cruz River. Because of the proximity of water for livestock in the Santa Cruz this entire area has been, and still is, grazed heavily by domestic livestock, largely cattle. This has worked to the detriment of the grasses and to the benefit of the less palatable mesquite, the seeds of which are spread by the grazing animals.

This area has been changing for many years from an open grassland to a savanna type with velvet mesquite as the dominant woody species, a change that is prevalent in extensive portions of southern Arizona. The underlying causes will be discussed later under *Discussion*.

Rangelands of this sort that have not been altered by many years of overgrazing and/or competition for moisture by mesquite normally support a wide variety of perennial grasses, and it may be assumed that this was the case at the time these monuments were erected. During the intervening years, however, the grazing pressure, which had begun shortly prior to the 1890s, has continued. As a consequence, grass density is now low and grass species are few. Only four species were recorded in 1983: blue grama, sideoats grama, curly mesquite, and threeawn.

A variable-aged stand of velvet mesquite, indicating continuing invasion, constituted the principal woody species here in 1983. Two other scrub species, huajilla and banana yucca, occurred occasionally.

Monument No. 116. Semidesert Grassland

MONUMENT NO. 117. Semidesert Grassland

Monument No. 118. Riparian

Monument 118 was constructed of masonry from the remains of Emory's Marker No. 25, in the floodplain of the Santa Cruz River, only a few feet east of the main channel. The river is subject to periodic heavy flooding and channel shifting, with the apparent result that the 1893 masonry monument was washed away or destroyed. It has been replaced by a standard iron obelisk about seventy-five feet east of the original. My repeat photograph shows the approximate location of the masonry marker and was taken facing west, as in the 1893 photo.

When No. 118 was built, the river banks and adjacent floodplain were comparatively free of woody vegetation. Most of the trees or shrubs that were growing at the time were small. Today, by contrast, the floodplain supports a dense forest of trees and shrubs with an almost complete ground cover of annual forbs, belonging primarily to the sunflower and mustard families. Most of the trees are black willow, with only an occasional Fremont cottonwood. The large tree that fills the left quarter of my repeat picture is cottonwood; the rest are willows. The shape and general appearance of the central tree behind the monument in the earlier picture suggests that it, too, was cottonwood.

The apparent major vegetational change here, in addition to the differences in woody vegetation, seems to be replacement of perennial forbs and grasses by annual forbs. Prior to the time the monument was built, this floodplain probably supported a dense stand of sacaton. This is suggested by the dark color of the soil, by the level plain with groundwater near the surface, and by the scattered remnants of a grass that appears to be sacaton that show in the 1893 picture.

MONUMENT NO. 118. Riparian

Monument No. 119. Evergreen Woodland, Semidesert Grassland

This marker was erected in the rolling hills about midway between the Santa Cruz River and the town of Nogales. The vegetation here today, as in 1893, is primarily grasses with a scattered stand of low-growing trees. The trees in the earlier picture appear to be largely, or entirely, oaks. Today the oaks, both Mexican blue and Emory, remain, but are interspersed with variable-aged velvet mesquite, indicating past and continuing invasion by this aggressive species.

The monument was erected in June 1893 prior to the onset of the summer rains, and the early picture shows the effects of the preceding two years of extreme drought. Vegetation in the area near the monument also seems to have been destroyed during the monument-erection operation.

When I took the repeat picture in August 1983, I recorded five perennial grasses: slender grama, spruce-top grama, threeawn, curly mesquite, and cane beardgrass and an abundance of huajilla and scattered plants of the tuberous-rooted forb amoreuxia.

There has been no change in life form here, the vegetation remaining a largely open grassland, but one that is in the process of being converted into savanna by the invading mesquite.

MONUMENT NO. 119. Evergreen Woodland, Semidesert Grassland

Monument No. 120. Semidesert Grassland

The area showing in the 1893 picture appears to have been burned not long before the monument was established. The proximity to Nogales, with the consequent possibility of incendiarism, supports this conclusion. The woven-wire fence that now separates Mexico and the United States in this area obstructs the view but not the entry into the U.S. of illegal aliens. At the time the picture was taken, the fence opposite the monument had been cut from top to bottom, leaving a gap through which a horse could have been ridden.

As I have commented at other monuments east of here, but that applies most particularly to the upland area west of the Patagonia Mountains, the former grassland is changing to a savanna type, with velvet mesquite as the tree species. And here, as generally, this is an ongoing process, as indicated by the variable-aged stands of the mesquite.

I recorded only three woody species here, primarily velvet mesquite, but with some low-growing and closely grazed huajilla and an occasional sotol. Note the defunct sotol to the left of the monument in the original picture, and the living individual at the extreme right of my repeat photo.

I recorded only two perennial grasses as common here in 1983: sideoats grama and spruce-top grama. The entire area supported an abundance of annual goldeneye, a forb that indicates range deterioration—that is, a breakdown of the perennial grass cover.

Monument No. 120. Semidesert Grassland

Monument Nos. 121 and 122. Urban

Both of these boundary markers lie within the municipal limits of the twin border cities of Nogales, Arizona, and Nogales, Sonora. Both have grown from the sprawling, thinly populated communities that they represented in the later years of the nineteenth century, to the cities they are today. Nogales, Arizona's population at the time of the 1980 census was 15,700; that of Sonora, Mexico for the same year was 1.5 million.

Monument 121 overlooks the cities from the top of a steep hill immediately to their east; No. 122 is on a street of Nogales, Sonora. The only vegetation at No. 121 is a few plants of Russian thistle. These have replaced a few perennial grasses and a little sotol that were growing there in 1893. At No. 122 the former cottonwoods have vanished and been replaced by parked automobiles.

MONUMENT NO. 121. Urban

Monument No. 122. Urban

MONUMENT NO. 123. Semiurban, Semidesert Grassland

This marker was placed on a small hill in an area that was open grassland in 1893. Today, new apartment building construction and a Mexican trash dump extend almost to the monument, and a recently completed border entry gate lies in the drainage about a hundred yards to the west.

Despite these man-made changes, much of the native vegetation still remained in the vicinity of the monument when I photographed it in August 1983. Four woody species were common at that time: velvet mesquite, wait-a-minute bush, huajilla, and velvet-pod mimosa. Interspersed between the individuals of these scrub species were four perennial grasses: slender grama, spruce-top grama, sideoats grama, and threeawn. The area appeared not to have been grazed for some time, which may explain the occurrence of these palatable grasses in the midst of civilization. These same species, both scrub and grass, probably also occurred here at the time the monument was erected.

The trees on the hills in the background are oaks and mesquite.

Monument No. 123. Semiurban, Semidesert Grassland

Monument Nos. 124 and 125. Evergreen Woodland

Senate Document 247 (p. 190) describes the area west of Nogales in the Pajarito Mountains as difficult of access:

> Beyond Nogales the country gradually rises, and on approaching the Pajarito Mountains becomes very rough and broken by deep, precipitous ravines; wagon transportation was impossible, and resort to pack animals was necessary.

Monuments 124 and 125 are located in this gradually rising area, and I was able to reach them after some walking, without great difficulty. Both are in a region characterized by an overstory of oaks and an understory of perennial grasses and shrubs.

The two oaks most common in southern Arizona at elevations between 3,000 and 6,000 feet, Mexican blue and Emory, occur here, with some velvet mesquite beginning to invade at No. 124. Although grasses predominate beneath and between the trees, shrubs are also common.

The area is grazed by cattle, rather heavily on the Mexican side of the line, fairly conservatively on the American, a degree of use that has permitted the survival of a variety of palatable perennial grasses. The most common of these in 1983 were sideoats grama, slender grama, spruce-top grama, plains lovegrass, sprangletop, and cane beardgrass. The principal woody species, in addition to the two oaks, were wait-a-minute bush, huajilla, velvet-pod mimosa, and sotol.

Although there has been a marked change in the appearance of the vegetation with the passing of time at these monuments, the principal life forms have remained the same. At Monument 124, the oak to the right of the monument in both pictures is probably the same individual but showing considerably more growth in the later picture. The tree visible on the far left of the earlier picture has disappeared during the ninety-year interim. At Monument 125, the stand of oaks that appears to have been rather mature in 1893 has entirely vanished and the trees are beginning to be replaced by younger individuals.

Monument No. 124. Evergreen Woodland

MONUMENT NO. 125. Evergreen Woodland

Monument No. 126. Evergreen Woodland

This monument was erected at one of the highest points (5,300 feet) along the entire border. The vegetation was then, and still is, evergreen woodland with a grass and shrub understory. The lone oak behind the monument in the original picture has now disappeared, and the complete area has been taken over by a jungle of trees where formerly there had been bare ground. At the time of my visit in November 1983, the monument would have been almost entirely hidden had I not broken and removed numerous branches before taking the repeat photograph. This increase in growth characterized most of the immediate area.

The ground in the vicinity of this monument now supports a dense stand of vegetation that consists of a greater number of trees and coarse-growing shrubs than at the two previous monuments. In addition to the Emory and Mexican blue oaks that were growing at Monuments 124 and 125, alligator juniper and Mexican manzanita complete the overstory here, with wait-a-minute bush, squawbush, and sotol making up the woody undergrowth.

No grasses can be seen in the original picture, probably because of soil disturbance while the monument was being erected. When I took the repeat photo, by contrast, five perennial grasses were common: plains lovegrass, wolftail, sideoats grama, hairy grama, and bullgrass. There is no reason to suppose that these same species would not also have been characteristic of the general area ninety years earlier.

Monument No. 126. Evergreen Woodland

Monument No. 127. Evergreen Woodland

This monument marks the western end of the boundary along parallel 31°20′ at the point where the International Line inclines slightly to the north before again changing direction at the Colorado River, 234 miles and 78 monuments to the west. Monument 127 is of masonry, stands at the site of the original Monument No. 27, and was constructed from the remains of that monument.

Senate Document 247 (p. 190) notes with reference to the boundary along parallel 31°20′:

> Exclusive of No. 53 at its eastern extremity, this section of the boundary is marked by 74 monuments; 13 are of masonry, located on sites of old monuments; 47 are new solid iron, and 14 new sectional iron. . . . The whole time consumed in their erection was from the 16th of June until September 19, 1893, a period of eighty working days, being an average of six days for each stone monument and for each iron monument one and three-tenths days.

There were few or no roads in the area in the 1890s, and erection of the monuments proved to be arduous. Again, as indicated in Senate Document 247:

> This range [the Pajarito Mountains] is so broken and cut up by deep, precipitous canyons that the work of erecting monuments was attended with more difficulties than were met in the same distance upon any other part of the boundary.

Although the rough terrain made it difficult for me to locate some of the monuments, there are now roads of sorts, or jeep trails, that enabled me to reach even the most inaccessible by walking or on horseback.

As at Monument 126, although the general vegetation today remains much as it was ninety years ago, there has been a great increase in the size and density of the woody vegetation near Monument 127. Even though no tree stumps show in the foreground of the original picture, some clearing must have been required for its construction. Despite this possibility, there seems to have been a large increase in tree size and total cover during the years that have intervened. My field notes here, as recorded initially on tape, read: "I am currently at Monument 127, which is located in a jungle, mostly of manzanita." As may be seen from the repeat photo, this "jungle" almost totally obscured the monument. A comparison of the two pictures indicates that the oaks and manzanita on the hill in the background have also similarly increased.

The principal woody species here when I took the repeat photo were Mexican manzanita, Mexican blue oak, silktassel bush, and squawbush. These formed so dense a cover as largely to preclude the establishment of a ground cover of grasses or other vegetation. The largest and most abundant (and aggressive) of the nonwoody plants was that nongrass member of the lily family, beargrass, pretty well holding its own with the larger trees and shrubs. Only two grasses were noted, bullgrass and plains lovegrass.

MONUMENT NO. 127. Evergreen Woodland

Monument No. 128. Evergreen Woodland

Monument 128 was placed on a high ridge representing the highest point along the boundary in the Pajarito Mountains, 5,453 feet above sea level. The monument was erected only after considerable difficulty (Senate Document 247):

> The pieces for this monument and the cement for its base were carried on pack mules 22 miles over very difficult mountain trails. The water for concrete was carried 9 miles and the sand 2 miles.

Fortunately, a jeep trail, very steep and washed out in places, and often barely navigable for our VW camper, enabled us to drive to within about a quarter mile of the site.

Although there have been drastic changes in the individual plants here, the overall vegetation appears to have changed very little over the years. No sign remains of the tree stubs showing in the original picture. Both of these were apparently oaks.

Trees and large shrubs are dominant here today, with a rather diverse grass understory. Mexican blue oak and alligator juniper are the principal trees at the monument; silktassel bush and Mexican manzanita are the dominant large shrubs. Between these grows a scattering of the insistent wait-a-minute bush with its paired, recurving thorns, sotol, beargrass, and banana yucca.

Because the trees and shrubs form a less dense cover than at Monument 127, there is growing space here for a number of grasses, chiefly bullgrass, plains lovegrass, Texas beardgrass, wolfspike, wolftail, and hairy grama.

Lacking evidence to the contrary, it may be assumed that all of these same species characterized this area when the monument was established.

MONUMENT NO. 128. Evergreen Woodland

Monument No. 129. Evergreen Woodland

Although Senate Document 247 says only that the sand, cement, and water for constructing this monument had to be packed for several miles, it also proved to be difficult for me to reach. At one point on a steeply sloping side hill our car was in imminent danger of tipping over and rolling down into a deep canyon. In addition, the walk to the monument and back was arduous, requiring hours of climbing up and down steep, rocky slopes.

This masonry monument was constructed on the site, and presumably from the remains of, original Monument No. 18. When I visited the area in 1983, a large alligator juniper occupied the original camera station, making it impossible to duplicate the former view of the monument. The initial picture was taken facing very slightly north of east; my repeat shows the view looking north into Arizona.

There has been a change in the vegetation here over the years, although perhaps not as great as the photographs seem to indicate. The original has too little contrast to show clearly the woody vegetation. Despite this, however, the general area today almost certainly does support more trees than it did ninety years ago, a probable result of the U.S. fire-control program that has been in effect most of this time. The hills in the background of the original picture, although they may appear to be largely devoid of vegetation, would have supported a good ground cover of grasses when the monument was constructed, in addition to any trees and shrubs.

In November 1983, this entire area was vegetated with an open stand of alligator juniper, Mexican blue oak, squawbush, wait-a-minute bush, sotol, and beargrass. These were accompanied by a rich grass flora consisting principally of plains lovegrass, sideoats grama, hairy grama, wolftail, Texas beardgrass, and cane beardgrass.

The rugged terrain, water developments for cattle, and abundant forage, make this an excellent habitat for white-tailed deer as was evidenced at the time of my visit by the frequence of their tracks and feces.

Monument No. 129. Evergreen Woodland

Monument Nos. 130 and 131. Semidesert Grassland

These monuments lie at an elevation above sea level somewhat lower than No. 129. The resultant slight increase in aridity is enough to change the general vegetation aspect from woodland with a grass understory to grassland with the trees restricted largely to the north-facing slopes.

The pictures provide no evidence of any material change during the last ninety years in the kinds of vegetation that characterized the areas around either of these monuments. The photograph of No. 130, which is on a north-facing slope, showed a few trees in 1893, and still does today; that of No. 131, on a steep south slope, is treeless in both pictures. Grass at both sites is much more abundant today than formerly, probably due in part to the two years of severe drought immediately prior to erection of the monuments, and the heavy grazing to which the entire region had been subjected.

At Monument 130, sotol and a few trees (probably oaks) can be distinguished in the early photo, as they still can today. Emory oak is the only tree species growing at the site today. In addition, a number of other woody plants or halfshrubs occur occasionally. These are wait-a-minute bush, hopbush, lemonade berry, velvet-pod mimosa, kidneywood, range ratany, and wooly sagebrush. Despite the rather large number of shrubs, grasses are by far the most abundant life form and dominate the landscape. The most abundant grasses at Monument 130 at the time of my visit were cane beardgrass, sideoats grama, plains lovegrass, wolftail, and sprangletop.

The hillside in the vicinity of Monument 131 supports a different variety of scrub vegetation than the habitat of the preceding monument, apparently reflecting the greater aridity of the south-facing slope. In addition to velvet mesquite that is beginning to invade, I recorded ocotillo, sotol, yerba de pasmo, huajilla, trixis, Parry sage, twinflower, and limberbush. Grasses were abundant, but consisted of only a few species—sideoats grama, cane beardgrass, threeawn, and a member of the genus *Hilaria.*

Here, as at Monument 130, deer signs were plentiful, and I encountered one white-tailed doe and, in addition, a herd of javelinas.

MONUMENT NO. 130. Semidesert Grassland

MONUMENT NO. 131. Semidesert Grassland

Monument Nos. 132–35. Semidesert Grassland

These four monuments, although varying somewhat in the plant species found in the vicinity of each, are all in the same major ecosystem and have enough species in common to permit them to be discussed as a unit.

No. 132 lies on a hilltop in a remote area far from any roads, and I reached it only after three hours of hiking across rough, but interesting terrain, pock-marked by old, abandoned gold and silver mines.

The original picture here shows open grassland totally devoid of any woody vegetation. This portion of the boundary is still predominantly grassland, but with a scattering of shrubs and varying numbers of mesquites. Velvet mesquite is the only tree growing in the vicinity of Monument 132, but four shrubs—ocotillo, huajilla, velvet-pod mimosa, Gregg dalea, and snakeweed—are common. Although close grazing made grass identification difficult, I was able to distinguish four species: wolftail, threeawn, slender grama, and sideoats grama.

Monuments 133–35 are all on the long-established Tres Bellotas Ranch. Because a Mexican bandido and alleged murderer was at large in the area, we set up camp at the ranch headquarters, and the ranch owner, Lyle Robinson, and I rode to the monuments on horseback.

A mixture of grasses and shrubs characterized the area around Monument 133 when it was erected, and still does today. Three shrubs that appear to be oaks and a single sotol plant show in the original picture. Both of these species still grow here today, and both can be seen in the repeat photo. In strong contrast with the vegetation shown in the original picture, however, ocotillo is the principal woody plant today. Both of the common oaks, Mexican blue and Emory, as well as velvet-pod mimosa, sotol, prickly pear, and Russian thistle are also abundant here. Accompanying these are several kinds of grasses, principally Arizona muhly, wolftail, plains lovegrass, sideoats grama, spruce-top grama, and threeawn. Despite the greater number of tree and shrub species, and the relatively few grass species, the total number of woody plants is far outnumbered by the grasses, giving the area an overall grassland aspect.

The area around Monument 134 appears as open grassland, devoid of trees or shrubs in the 1893 photo. Today, although still basically grassland, it is no longer open but is being vigorously invaded by velvet mesquite. Several young, vigorous trees show in the repeat photo. Other mesquites in the general area vary in size, indicating current, active invasion.

In addition to the mesquite, other common woody species are Mexican blue oak and algerita in the draws and on north-facing slopes, with velvet-pod mimosa, huajilla, range ratany, jumping cholla, and cane cholla occurring more widely. The principal accompanying grasses are slender grama, sideoats grama, curly mesquite, cane beardgrass, and threeawn. Although ephemeral forbs depend on seasonal rainfall and may be present one year but not the next, both annual goldeneye and plantain were abundant in the area in late January 1984.

Except for a few minor differences in species, the area around Monument 135 is essentially similar to that at the preceding three monuments. Although at a similar elevation where oaks might be expected, there were none in the immediate vicinity. The only tree visible in the earlier picture is probably oak but may be mesquite; those showing left of the monument in the repeat photo are all mesquite. Ocotillo and huajilla are the most common shrubs. The perennial grasses are essentially the same as at Monument 134.

Monument No. 132. Semidesert Grassland

MONUMENT NO. 133. Semidesert Grassland

Monument No. 134. Semidesert Grassland

Monument No. 135. Semidesert Grassland

Monument No. 136. Semidesert Grassland

Monument 136 was constructed at the site of old No. 17, of "dressed stone," presumably from the remains of the old marker. "The cement, water, and sand were carried several miles on pack mules" (Senate Document 247, p. 191). The monument stands 4,514 feet above sea level in a slight saddle near the summit of a disjunct mountain mass, Cerro del Fresnal.

This is basically semidesert grassland but with an atypically large number of scrub species. The vegetation here seems to have changed little during the past ninety years.

Mexican blue oak is found in places on the slopes of the mountain, but there is none today on the summit. Numerous other woody taxa, however, are common on the mountain top in the vicinity of the monument. These are principally desert scrub oak, hopbush, sotol, yerba de pasmo, turpentine bush, range ratany, ocotillo, and Gregg dalea. A single globose member of the cactus family, rainbow cactus, is fairly common. Accompanying these are several perennial grasses, largely sideoats grama, hairy grama, plains lovegrass, wolftail, and curly mesquite.

The steep slopes of Cerro Fresnal tend to restrict cattle access to the higher portions. This has resulted in moderate to light grazing pressure and consequent maintenance of the climax vegetation.

Monument No. 136. Semidesert Grassland

Monument No. 137. Semidesert Grassland

This, another masonry monument, was built at the site of old No. 18.

No trees or shrubs can be distinguished in the original photo; in contrast, they occur commonly throughout the area today, and many show in the repeat picture. The former open grassland is changing here, as in so much of south-central Arizona, into a vegetation type dominated increasingly by mesquite and other scrub taxa. Some scrub species would have been intermixed with the grasses when the earlier picture was taken but were probably much less common than they are today. And mesquite, the most conspicuous and common woody plant today, did not then have the widespread distribution it now enjoys.

In addition to velvet mesquite, other woody species that are now common here are ocotillo (shown in the lower corners of the repeat photo), velvet-pod mimosa, Gregg dalea, and huajilla. Three globose cacti are also common: barrel cactus, rainbow cactus, and an unidentified species of hedgehog cactus.

Despite the increase in scrub vegetation, the grass flora in the vicinity of Monument 137 is richer and provides a markedly greater ground cover than the scrub. When I photographed the monument in late January 1984, I recorded nine perennial grasses: sideoats, slender, spruce-top and hairy gramas; plains lovegrass; cane beardgrass; wolftail; Arizona muhly; a perennial threeawn; as well as one annual, six-week threeawn. The range here is lightly grazed, a degree of use that has benefitted the grasses.

Monument No. 137. Semidesert Grassland

MONUMENT NOS. 138 AND 139. Semidesert Grassland, Sonoran Savanna Grassland

These monuments lie 2.5 and .2 miles, respectively, east of the small border-crossing town of Sasabe. The old Garcia Ranch buildings show to the immediate right of Monument 138; a portion of Sasabe can be seen slightly farther to the right in the far distance.

There has been a major vegetational change throughout this area during the last ninety years, a change evidenced primarily by the invasion of mesquite, cacti, and other shrubs and halfshrubs. At the time the monuments were erected, this was open grassland free of scrub vegetation and with mesquite restricted to the occasional drainage. When I first passed through here as a graduate student in 1930, it was still essentially open, rolling prairie grassland.

A large-scale, extensive mesquite eradication program has been carried out recently north of the boundary here on what is now known as the Buenos Aires Ranch. Were it not for this program, the area showing to the right of Monument 138 and to the left of No. 139 would have appeared in the repeat photos as a mesquite woodland, as it still is across the line in Mexico. The range, both north and south of the boundary throughout this area, has been heavily grazed for many years, and grazing pressure is still extreme on the Mexican side. In Arizona, not only has much of the mesquite been removed, but grazing pressure has been reduced and extensive areas have been reseeded with two exotic grasses from South Africa, Boer and Lehmann lovegrass. These attempts at range restoration, however, have not restored the land to its original pristine condition, and the mesquite and a variety of other scrub species are rapidly reinvading.

Excessively heavy grazing south of the border here, combined with the range restoration practices on the north, have combined to result in rather distinct vegetational differences on the two sides of the line. Thus, in Mexico the dominant woody plant today is velvet mesquite, with a scattering of ocotillo, velvet-pod mimosa, cane cholla, and prickly pear and an abundance of those indicators of a breakdown in perennial grasses—snakeweed and burroweed. There is none of the highly palatable huajilla, nor are there any perennial grasses in Mexico at Monument 138. At No. 139 also there is no huajilla and only a little sideoats grama, slender grama, and threeawn.

Across the boundary fence in the United States, under the Buenos Aires Ranch program of range improvement (as contrasted with Mexico), there is a marked difference in both woody plants and grasses. Most of the mature mesquites have been eradicated by bulldozing but are beginning to reinvade. Both snakeweed and burroweed are as abundant as they are across the line, but huajilla, absent to the south, is moderately abundant on the north. As a result of the reseeding program both Boer lovegrass and Lehmann lovegrass are everywhere abundant. At Monument 139 these grasses are combined with the native sideoats and slender gramas, and with threeawn.

MONUMENT NO. 138. Semidesert Grassland, Sonoran Savanna Grassland

Monument No. 139. Semidesert Grassland, Sonoran Savanna Grassland

Monument No. 140. Semidesert Grassland

Monument 140, at an elevation of 4,028 feet above sea level, although located for the most part in the semidesert grassland, lies at the lower edge of the evergreen woodland. As a consequence, the vegetation near the monument contains some characteristic woodland species.

At the time of my visit in February 1984, this area was heavily grazed by cattle, both north and south of the boundary fence, and appeared to have been so for many years. Long-continued grazing pressure has favored the woody plants, many of which have little or no value as feed for cattle, and these have tended to replace the relatively palatable grasses as they have been weakened or killed by overgrazing.

A comparison of the two photographs reveals no change in life form here over the past ninety years. There are some apparent taxonomic differences but these may be due, at least in part, to a lack of clarity and detail in the original.

Although there are many grasses growing here today, woody plants show more prominently in the picture and appear to predominate. Most common among these are ocotillo, sotol, cane cholla, huajilla, velvet mesquite, velvet-pod mimosa, limberbush, trixis, Arizona rosewood, and snakeweed. Barrel cactus occurs occasionally. Mexican blue oak and rosewood can be seen on the near skyline as strays from the higher-lying woodland.

Two perennial forbs, ragweed and Parry penstemon, occur as scattered individuals; an ephemeral forb, annual goldeneye, was abundant at the time the repeat picture was taken.

Protected in large part by the rocks and bushes, are a number of grasses. The most common of these are cane beardgrass, threeawn, slender grama, sideoats grama, and plains lovegrass.

Monument No. 140. Semidesert Grassland

MONUMENT NO. 141. Semidesert Grassland

Monument 141 is described in Senate Document 247, p. 191, as "occupying a commanding position on the crest of the Pozo Verde Mountains, . . . a masonry monument built on the site of old No. 15. The difficulties here were such that the cement and water were carried by pack mules $2\frac{1}{2}$ miles." This latter statement I was able to appreciate after walking to the monument across deep canyons and up high, steep, and rocky hills. Because of the difficult terrain the area remains today almost ungrazed by cattle on the Mexican side and only lightly grazed on the American.

The original picture shows too little of the surrounding area to be of much value in determining possible changes in vegetation. The plants that do show suggest little or no change. Sotol, with three individuals visible in the original, is still common there today. Other common woody plants now growing there are ocotillo, velvet-pod mimosa, kidneywood, Gregg dalea, two halfshrubs—snakeweed and burroweed, and one cactus—purple prickly pear. The principal grasses are cane beardgrass and curly mesquite.

MONUMENT NO. 141. Semidesert Grassland

Monument No. 142. Sonoran Desert

This, the easternmost monument on the boundary between Mexico and the Tohono O'odham Indian Reservation, lies slightly east, and within sight of, the old Buenos Aires Ranch headquarters. This is not to be confused with the ranch of the same name at Monuments 138 and 139.

The dominant vegetation at this site in 1893 was scrub, as it still is today. There were some grasses at that time, and there still are. Various woody species, none of which can be identified in the original picture, but which show as the dark background, largely obscured the grasses which show as light-colored areas in the center of the photograph.

At the time of my retake, the same general condition obtained, the woody plants and grasses both consisting largely of Sonoran Desert species, but with a few representatives of the generally higher-lying semidesert grassland to the east.

As will be noted in most of the descriptions that follow, the Sonoran Desert is dominated by scrub, the grass and perennial forb flora having a much lower density and biomass than they do to the east.

The principal woody vegetation in 1984 consisted of foothill paloverde (the large tree to the right of the monument), ocotillo, velvet mesquite (partially shown at the left margin of the picture), velvet-pod mimosa, limberbush, Parry dalea, range ratany, catclaw, jumping cholla, cane cholla, hedgehog cactus, and desert hackberry. The few grasses were represented by only three species: cane beardgrass, threeawn, and slender grama.

MONUMENT NO. 142. Sonoran Desert

MONUMENT NOS. 143 AND 144. Sonoran Desert

Monument No. 143 (142A) lies in the very extensive Baboquivari Valley, west of the Baboquivari Mountain range. The old obelisk has vanished and the site is now marked by 142A. The original concrete base remains, but this has had a more recent monument superimposed upon it. The 1979 U.S. Geological Survey San Miguel Quadrangle map shows 142A but no 143, supporting the assumption that the two are equivalent. Monument 144 is near an unauthorized border crossing (The Gate) which, at the time of this study, was operating as an entry point for illegal aliens and, presumably, drug runners. Although we camped for two nights about a half mile to the east, no one disturbed us.

There has been a drastic change in the plant cover at both of these monuments in the past ninety years. This portion of the boundary was primarily grassland in 1893 but has changed since then to mixed scrub with few or no perennial grasses.

The few trees that show at Monument 143 in the original picture are probably mesquites; a few may be paloverdes. When I took the repeat photo the only woody vegetation consisted of one tree species—velvet mesquite, one cactus—jumping cholla, and two halfshrubs—burroweed and snakeweed. There were no perennial grasses, but a single annual species—six-week threeawn—was common. All of these are so-called invaders, typical of run-down, deteriorated grasslands.

The same situation prevails at Monument 144. The former open grassland, free of scrub vegetation except for the long, dark streak in the background indicating a wash or drainage, has been replaced by scrub. Velvet mesquite, catclaw, and jumping cholla now dominate the scrub overstory, with snakeweed and burroweed as an understory. The only perennial grass is an occasional threeawn. Exceptionally heavy rains during the summer of 1983 stimulated a heavy growth of six-week threeawn. This shows as the light-hued ground cover in the background of the recent picture.

Monument No. 143. Sonoran Desert

Monument No. 144. Sonoran Desert

Monument No. 145. Sonoran Desert

Monument 145 stands in a severely eroded and gullied bajada in the western portion of the Baboquivari Valley. Although no boundary fence shows in the repeat photo, there is a well-constructed, five-strand barbed-wire fence about fifty feet to the north.

Senate Document 247 (p. 191) says, with reference to the Baboquivari Valley from about the Buenos Aires Ranch (Monument 142) to and including Monument 146:

> This is an arid region, though occasional rains occur, and considerable grass is found on some of the plains, which, with the aid of a few pumping wells, supports a number of herds of horses and cattle.

This entire area is still heavily grazed both north and south of the border, resulting in severe erosion and vegetation that contains few palatable forage plants. The gully that shows to the immediate right of the monument is three to four feet deep and has almost undercut the concrete base of the monument. At the time of my photograph, the gully was filled almost to the top with dead Russian thistle plants that had blown in. When the earlier picture was taken, there was no erosion.

In addition to the erosion, a comparison of the two pictures also shows marked changes in the vegetation. There was an abundance of mesquite with some grasses, probably six-week threeawn and/or Rothrock grama in 1893. Today the mesquite is still common, but most of the vegetation now consists of burroweed and snakeweed. There were no perennial, or even annual, grasses at the site when I visited it, the only herbaceous vegetation being three annual forbs—Russian thistle, groundsel, and mustard. Every species growing near the monument today is an indicator of overgrazing and range depletion. The resultant extensive soil loss reflects this abuse.

Monument No. 145. Sonoran Desert

MONUMENT NO. 146. Sonoran Desert

Boundary Monument 146 was built on the highest boundary crossing of Morena Mountain at the site of Emory's original No. 14. Although Senate Document 247 does not so indicate, it is probable that the stones used in constructing the present monument were those that had been piled up to form the original. As the monuments could be seen for long distances here they were placed at very nearly the maximum previously agreed upon allowable distance apart. Thus, No. 146 lies four miles west of 145, and No. 147 four and a half miles west of 146.

There has been little or no overall change in vegetation at this site since the monument was built. Ocotillos may be more abundant today than formerly; certainly some are now growing where none grew before. Sahuaros, also, may be somewhat more common today, with some individuals established and growing in places where none were growing in 1893. Despite these individual differences, however, the formerly dominant desert scrub still retains its dominance.

The principal scrub species that I noted in the early spring of 1984 were ocotillo, brittlebush, jojoba, limberbush, kidneywood, trixis, foothill paloverde, sahuaro, southwestern twinflower, huajilla, and tomatillo. There were small amounts of two perennial grasses—Rothrock grama and slender grama. Although these, and perhaps others, may be presumed to have grown here when the monument was built, they do not show in the picture.

Monument No. 146. Sonoran Desert

Monument No. 147. Sonoran Desert

Monument 147 stands in the extensive Vamori Valley eight miles southwest of the small Indian village of Vamori. When erected it was surrounded by an open creosotebush stand with grasses, probably Rothrock grama in part, growing between and beneath the bushes. Today, the once-open scrub is now a dense, closed community, with almost no grasses, either annual or perennial.

Creosotebush is strongly dominant here, but careful search revealed scattered individuals of several other scrub species—sahuaro, desert zinnia, triangle bursage, Mormon tea, tomatillo, ocotillo, velvet mesquite, and foothill paloverde. My initial impression was that there were no perennial grasses, but a more thorough search showed bush muhly growing beneath many of the larger creosotebushes where it had been protected from grazing. The only other perennial grasses were the almost entirely nonpalatable fluffgrass and an occasional threeawn. The former partial ground cover of grasses no longer exists.

Monument No. 147. Sonoran Desert

Monument No. 148. Sonoran Desert

Monument 148 stands on a gently sloping bajada near the eastern edge of Tecolote Valley. At the time of its erection the surrounding area was characterized by an open forest of mesquite with an understory of shrubs or grasses. Senate Document 247 says nothing about the vegetation of this area, nor does the original monument picture show the vegetation clearly enough to be of much help in determining the understory life form. The abundant mesquite, growing on a bajada, rather than on a floodplain site and the absence of any identifiable bursage in the original picture suggest the area as a recently (in 1893) deteriorated, former grassland. Based on this assumption, I postulate that subsequent grazing pressure has prevented revegetation by grasses but has permitted invasion by the totally unpalatable bursage.

Today, the overstory vegetation consists, much as it did in 1893, of an open forest of velvet mesquite. The ground between the trees now supports an almost pure stand of triangle bursage. Scattered here and there between the bursage bushes are infrequent plants of jumping cholla, four-wing saltbush, desert hackberry, and snakeweed. The only perennial grass that grows there now is bush muhly, and this occurs only where the mesquite or shrubs afford some protection from livestock grazing.

Admittedly, the assumption that there has been a change—or lack of change—in the vegetation here is based in part on my own experience, and in part on what the early picture reveals. The mesquite, however, in a situation of this sort strongly suggests the kind of successional change I have proposed.

I am, therefore, proposing the probability of a change here from a former grassland (prior to 1893), to mesquite with grasses, bare soil, and some scrub (in 1893), to mesquite and bursage today.

MONUMENT NO. 148. Sonoran Desert

Monument No. 149. Sonoran Desert

Monuments 148, 149, and 150 were not accessible with our VW camper and could be reached only after several miles of walking. I finally found No. 149 in Tecolote Valley, south of the boundary fence and about a third of a mile west of Chukut Kuk Wash (U.S. Geological Survey La Lesna Mts. Quadrangle Map, 1963).

Here, as at Monument 148, there has been a marked change in the vegetation during the past ninety years. Various scrub species dominated the scene in 1893, as they do today, but the genetic composition seems to have changed greatly. Paloverdes, if present, are difficult or impossible to distinguish in the earlier picture. The tallest tree that breaks the skyline to the right of the monument was probably an ironwood, in all likelihood the same one that grows in the same place today. The large shrub near the right edge of the original picture appears to have been creosotebush, again probably the same plant that stands to the right of the monument and in front of a much taller paloverde today. Large clumps of what may be bush muhly show to the left of the monument.

Foothill paloverde is dominant at this site today, and comprises most of the trees that show in the recent photo. In addition, there is some ironwood. A number of sahuaro cacti are also growing where few or none were visible previously. The large, branched sahuaro to the left of the monument, in particular, has become established and made a surprising growth during the ninety-year period.

The dominant understory shrub here, as at Monument 148, is triangle bursage. Growing with it are occasional plants of creosotebush, ocotillo, and pencil cholla. In addition, there are a few grasses: bush muhly, growing where protected by some of the shrubs, fluffgrass, and two threeawns—one an annual, the other a perennial.

MONUMENT NO. 149. Sonoran Desert

MONUMENT NO. 150. Sonoran Desert

Constructed of masonry on the floodplain of a large wash, Monument 150 stands on the site of original Marker No. 13. A high vertical bluff rises just south of the obelisk, as the wash swings in a wide S-shaped curve here before crossing into Mexico. This is a desert riparian site, and an accumulation of litter indicates periodic flooding from the nearby wash.

The original picture shows low-growing trees and other scrub vegetation in the foreground, with grasses and scrub in the background. In the recent photo the trees and brush have grown so large and so thick that the background is totally obscured. The brush was so dense at the time of my visit that I was unable to obtain an exact replica of the original picture, the former camera station having been taken over by a large mesquite.

The composition of the scrub species may have changed little since the monument was built, but the trees and shrubs have grown so large that the appearance of the site has been drastically changed. Only four species make up today's tangle of scrub—velvet mesquite, blue paloverde, creosotebush, and catclaw. Beneath and between them grows a heavy stand of grass, consisting primarily of Lehmann lovegrass, bush muhly, plains bristlegrass, and threeawn. Although the general area is grazed, the dense brush in the vicinity of the monument currently protects the immediate site from cattle and horses.

MONUMENT NO. 150. Sonoran Desert

Monument No. 151. Sonoran Desert

Monument 151 lies in Tecolote Valley one mile east of Tatkum Vo, a long-established Tohono O'odham Indian ranch. Cattle from the ranch range throughout the area and graze much of it heavily.

A comparison of the two pictures shows that there has been a marked change in the vegetation here during the last ninety years. A formerly open savanna with a good stand of grasses has been converted to scrub that is now almost totally devoid of grasses. Although positive identification of the kinds of grasses that show in the earlier picture is not possible, Rothrock grama probably predominated. Foothill paloverde and creosotebush appear to have been the principal scrub species.

Today the site supports the same two scrub taxa and, in addition, a great deal of triangle bursage. The increase or actual invasion of bursage where none showed before, not only at this monument but at 148 and 149 as well, is of considerable interest. This species has long been described as a representative member of the climax paloverde–creosotebush–bursage community. Yet here it functions as an apparent pioneer species, moving in to replace the former grasses. In this respect it is behaving like burroweed and snakeweed, both of which have long been recognized as so-called pioneers or invaders.

The only grasses present when I took the repeat photo were a little fluffgrass and very infrequent individuals of a perennial threeawn. The area both north and south of the boundary fence was heavily grazed, and the almost complete absence of any grasses would seem to indicate that this grazing pressure represented a condition of long duration.

MONUMENT NO. 151. Sonoran Desert

Monument No. 152. Sonoran Desert

The woody vegetation growing here in 1984 was rather similar to that at Monument 151. Lying as this area does, however, slightly more than two miles west of Tatkum Vo, the vegetation here is less heavily grazed, a fact that seems to be reflected in the current population of both grass and scrub.

Several scrub taxa that still characterize the area today can be distinguished in the 1893 photo. These include cane cholla, ocotillo, sahuaro, and triangle bursage. The grasses that formerly showed so prominently at the previous monument did not occur here. There were, with little doubt however, some grasses, but none can be distinguished with certainty in the photograph.

Today's woody vegetation at this site includes the four above-named species and, in addition, creosotebush, Englemann prickly pear, ironwood, range ratany, huajilla, jojoba, and tomatillo. Three palatable perennial grasses—bush muhly, slim tridens, and Arizona cottongrass—grow here in some of the bushes where protected from grazing animals. In addition, three of lesser palatability—fluffgrass, six-week schismus, and a perennial threeawn—are common between the bushes.

The richer flora here than at Monument 151 may be attributed in part to site differences, but must be due in considerable degree to lighter grazing pressure. Not only the grasses, but some of the woody species as well, notably huajilla and range ratany, are readily taken by cattle and are often killed under long-continued heavy grazing.

MONUMENT NO. 152. Sonoran Desert

Monument No. 153. Sonoran Desert

Monument 153 stands on a sharp peak near the southern end of the La Lesna Mountains. There was apparently no place on the narrow ridge to place a camera tripod, and the entire peak, with the monument barely visible against the skyline, was photographed from near the base.

Although a road of sorts shows in my repeat photo, it had been badly washed out in several places and was not navigable with our car.

This proved to be the most difficult of all the monuments to erect and is described in some detail in Senate Document 247:

> No. 153 . . . occupies the most remarkable position on the entire boundary. The line in crossing the Cerro de la Lesna rises abruptly from the plain below, a distance of about 500 feet, the upper 100 feet being a sheer precipice on both sides. To reach the summit of this ridge required a specially skillful and athletic climber to carry a rope, by means of which others were enabled to ascend and perform the work of erecting this monument. It is of the sectional iron type, the pieces and other materials being carried as far as possible on pack animals, and then hoisted by hand, with the aid of ropes, to the summit. The knife-edge crest was blasted off to give sufficient width for the base of the monument, which was then bolted to the solid rock. The erection of this monument proved to be the most difficult upon the entire boundary, requiring four days of excessive labor.

The original picture shows desert scrub with an understory of grasses. The grasses cannot be distinguished as to either genera or species, and only three scrub taxa can be identified—ocotillo, foothill paloverde, and sahuaro. When I took the repeat picture in late February 1984, triangle bursage was the most common scrub species. Other common scrub taxa were foothill paloverde, sahuaro, ocotillo, creosotebush, ironwood, and huajilla. A few perennial grasses occurred here and there, principally bush muhly and plains bristlegrass where protected by shrubs or trees, with threeawn and fluffgrass in the open. Here, as at the two previous monuments, there has been a marked increase in bursage and a reduction in grasses. The general scrub aspect remains, but the plants are larger and the cover is denser.

Monument No. 153. Sonoran Desert

Monument No. 154. Sonoran Desert

Monuments 154 through 158 lie in lower Kom Vo Valley, covering a distance of fifteen miles in a low-gradient, dusty, alluvial area shown on the U.S. Geological Survey Kom Vo Quadrangle map as "The Great Plain." Although the vegetation throughout this area is basically scrub, the taxa differ considerably from monument to monument, necessitating a separate discussion of each.

Senate Document 247 (p. 21) describes the vegetation throughout this general region, and including the area eastward from Lesna Mountain to Morena Mountain, as having "a luxuriant growth of mesquite, paloverde, palo fierro [ironwood] and cactus." The heavy grazing to which the area was subjected is suggested by the document statement that there was a Yaqui and Papago ranch a short distance west of La Lesna Mountain where "were seen large herds of cattle and horses."

Monument 154 was located in what was at that time a creosotebush flat with a good cover of grasses between the bushes. The grasses appear to have been either six-week threeawn, Rothrock grama, or a mixture of the two. Today, the creosotebush stand remains, intermixed with a little triangle bursage and an occasional velvet mesquite. The only grasses that have persisted are an occasional six-week threeawn and six-week schismus. When I visited the area in February 1984, the former grass cover had been replaced by annual forbs, principally a species of mallow, Texas filaree, combseed, a borage, lupine, and fiddleneck.

MONUMENT NO. 154. Sonoran Desert

Monument No. 155. Sonoran Desert

Monument 155 lies south of the Tohono O'odham irrigated fields known as Papago Farms. The soil on the boundary here is somewhat alkaline, as is indicated by some of the scrub vegetation.

There seem to have been only minor changes in the vegetation at this site during the past ninety years. Except for scattered velvet mesquites, today's plant cover consists almost entirely of two salt-tolerant shrubs—desert saltbush and threadleaf saltbush. A few mesquites show in the original picture as do numerous shrubs that were almost certainly the same saltbush species growing there today. Some perennial grasses may be seen in the left side of the 1893 photo, but none grow in the area today.

Monument No. 155. Sonoran Desert

Monument No. 156. Sonoran Desert

Three and one half miles west of Monument 155, No. 156 stands in what is today a cholla cactus–infested portion of the Great Plain. The original picture shows a rather open, nearly level plain with scattered shrubs, a partial ground cover of what appears to be grass, but no cacti.

Today, shrubs are much more abundant, and cholla cactus has become one of the principal species. Jumping cholla, shown prominently in the recent photo, is accompanied by lesser amounts of cane cholla. The other most common woody species are burroweed, tomatillo, creosotebush, desert saltbush, and velvet mesquite. There is a little six-week threeawn and Santa Rita threeawn, but no other grasses. A single perennial forb, canaigre, occurs rather commonly throughout the area. Burroweed, with its proclivity to move in where former semidesert-grassland grasses have been thinned out or destroyed, strongly suggests that this site at one time did, indeed, support grasses for the most part and that a change from grass to scrub was under way at the time the monument was erected.

Monument No. 156. Sonoran Desert

MONUMENT NO. 157. Sonoran Desert

Monument 157 is located on the west side of the extensive San Simon Wash that carries the runoff from Kom Vo and Chukut Kuk valleys. Here, as elsewhere generally in the Great Plain, a once open area has been taken over by scrub. Low-growing shrubs, probably saltbush or burroweed, show in the original picture, with a good carpet of what appears to be grasses or a mixture of grasses and annual forbs. Although two lines of darker vegetation that may have been mesquite show in the distant background, there was none evident in the vicinity of the monument.

Today, mesquite has become established throughout the area, and I even had to chop away part of one tree to photograph the monument from near the original camera location.

As the repeat photo indicates, the dominant woody plant now in the area is velvet mesquite. This is intermixed with lesser amounts of allthorn. Two halfshrubs—threadleaf saltbush and burroweed—are more numerous but less visible and with a considerably smaller biomass than the much taller and more heavily branched mesquite.

A small wash has developed a few feet north of the monument (hidden from view in the recent picture by the shrubbery), where there was none in 1893.

Monument No. 157. Sonoran Desert

Monument No. 158. Sonoran Desert

In 1893 this site was, as it still is, a creosotebush flat with trees bordering a wash showing in the background. The original picture shows some grasses between the bushes; there are no perennial grasses in the vicinity today.

The only woody plants occurring near the monument today are creosotebush and triangle bursage. At the time of the earlier photo some of the vegetation may have been bursage, but this cannot be determined with certainty. The trees showing in the background of the recent picture are blue paloverde, ironwood, and velvet mesquite. Paloverde and ironwood almost certainly also show in the 1893 photo; if mesquite was present it cannot be distinguished from the other trees.

No perennial grasses grow in the vicinity of the monument today. A single annual grass, six-week schismus, and two ephemeral forbs—plantain and an unidentified member of the sunflower family—were rather abundant at the time I took the repeat photo in early April 1984.

MONUMENT NO. 158. Sonoran Desert

MONUMENT NO. 159. Sonoran Desert

Monument 159 was placed on the crest of a malpais basalt mountain, 520 feet above the adjacent plain. The mountain bears no name on the Geological Survey maps but was considered by the monument crew to be an isolated outlier of Nariz Mountain lying about two miles to the west.

Except for individual plant location differences, there has been no visible major vegetational change here during the past ninety years. The paloverde that fills the right side of the original picture has now vanished, as have the ocotillos. There was no sign in the 1893 photo of a tree to the left of the monument where a mature paloverde now flourishes. The jojoba that shows with wide, light-colored leaves in the lower foreground of my recent picture seems to be the same plant as in the original, suggesting that this may be a long-lived species.

My notes of April 2, 1984, recorded the principal woody taxa here at that time as foothill paloverde, brittlebush, and jojoba, with range ratany and Pringle mercury second in abundance, and organpipe cactus, tomatillo, hopbush, and silverbush occurring occasionally. There was an abundance of six-week threeawn remaining after the previous summer's rains, and lesser amounts of Rothrock grama.

Because of the rocky terrain, steep slopes, and elevation above the plain, this area receives minimal to no use by domestic livestock. Man and his animals, therefore, appear not to have affected either the site or its biota.

Monument No. 159. Sonoran Desert

Monument No. 160. Sonoran Desert

Located on the highest point where the boundary crosses Sierra de la Nariz, this imposing masonry obelisk was constructed at the site of Emory's original No. 12. Standing, as it does, 1,000 feet above the bajadas surrounding the mountain, this monument provides a beacon of sorts that is visible for several miles, particularly from the east in the sunlight of early morning.

The steep slopes, large rocks, and rugged character of Sierra de la Nariz preclude the possibility of domestic livestock grazing here. As a consequence, the vegetation represents the climatic climax for the area, largely shrubs and one low-stature tree—foothill paloverde. The most abundant shrubs at the time of my visit were brittlebush, jojoba (shown partially obscuring the right corner of the monument), ocotillo, range ratany, Pringle mercury, and Sonoran croton. The only grasses were an occasional Arizona cottongrass and six-week threeawn.

In the original photograph brittlebush is easily identifiable in the lower left corner of the picture and perhaps also on the skyline adjacent to the monument. The other shrubs cannot be identified. As at Monument 159, there is no evidence of any change in the vegetation here during the period that has intervened between the two photographs.

Monument No. 160. Sonoran Desert

Monument No. 161. Sonoran Desert

This monument, of iron, stands in a creosotebush flat in Ali Chuk Valley, one mile south of the small Indian village bearing the same name.

The early photo shows no perennial vegetation except creosotebush but with a suggestion of what appears to be a sparse undercover of ephemerals. Today, the creosotebushes are still dominant but with an occasional jumping cholla and with ironwood and blue paloverde in a background wash. These same trees may have been present but indistinguishable in the earlier picture. Thus, the vegetation here remains today much as it was in 1893, a creosotebush flat.

Monument No. 161. Sonoran Desert

MONUMENT NO. 162. Sonoran Desert

This, another of the picturesque masonry obelisks, stands on a high ridge of the Santa Rosa Mountains (Sierra de Santa Rosa on U.S. Geological Survey, Diaz Peak Quadrangle map). This was built on the site of original Monument 10, and its construction "was attended with much difficulty, pack animals being necessary for transportation up the steep mountain slopes, and considerable work was expended upon the trail to make it passable for mules" (Senate Document 247, p. 192). The monument was built on a narrow ridge that drops steeply both to the north and south, on the south almost vertically for 75 to 100 feet.

I had some difficulty reaching the site carrying only a backpack with camera equipment and can appreciate the labor that must have been involved in building a trail up the boulder-strewn steep slope and transporting the sand, cement, water, tools, and other equipment that would have been required to construct the monument. I was able to pick my way between the large boulders or at times to jump from one to another; the mules might have been able to thread their way between in some instances. They certainly could not have taken the saltative route that I could.

This monument site, like the previous two, is inaccessible to domestic livestock. As a consequence, the vegetation has not been affected by grazing during the period since the monument was built. Even possible grazing by game animals is light, as I noted no signs of either deer or bighorn sheep en route to or in the vicinity of the monument.

A study of the first picture shows some half-dead or dead shrubs and, although none of them can be identified with certainty, their general appearance suggests that they may be either limberbush or torote. A comparison of the two photos indicates that there has been a marked increase in both trees and shrubs, and my field notes record a current floristic richness not visible in the original photograph.

Almost all the vegetation on the site today consists of either low-stature trees or shrubs. Brittlebush, hierba de la flecha, foothill paloverde, and creosotebush are most abundant; range ratany, tomatillo, trixis, southwestern twinflower, silverbush, ocotillo, sahuaro, organpipe cactus, hedgehog cactus, odora, and wedgeleaf limberbush occur less frequently, but are still common. The only grasses are two ephemerals—six-week threeawn and six-week schismus—and the only forb an annual unidentified legume.

MONUMENT NO. 162. Sonoran Desert

Monument No. 163. Sonoran Desert

Monument 163 stands on a western outlier of the Sierra de Santa Rosa in somewhat the same type of terrain (but more accessible) as Monument 162. This site also shows no signs of ever having been grazed by domestic livestock nor, at the time of my visit on March 2, 1984, did I see any indication of current use by deer or bighorn sheep.

A comparison of the two pictures gives no indication of any change in life form here. Individual plants have gone and come but the general life form remains today, as before, southern desert scrub. The cactus skeleton in the original photo is organpipe cactus; the bushy plant in the lower right-hand corner is probably foothill paloverde as also may be the two trees on the left of the picture. Brittlebush is identifiable as the low bush partially shown in the lower left corner. All of these species still grow in the area today.

In the repeat photo brittlebush shows as the common round-topped, white bushes; a sahuaro frames the picture on the right edge; the shrub outlined against the left skyline is creosotebush.

In my field notes I listed as most abundant brittlebush, wedgeleaf limberbush, range ratany, Sonoran croton, and hierba de la flecha. Occurring less frequently, but still common, were trixis, triangle bursage, organpipe cactus, sahuaro, hedgehog cactus, ocotillo, and southwestern twinflower. A single perennial grass—threeawn—occurred occasionally.

Monument No. 163. Sonoran Desert

Monument Nos. 164–66. Sonoran Desert

These three monuments lie in the Sonoyta Valley, west of Sierra de Santa Rosa and east of the Mexican village of Sonoyta. The boundary here separates the state of Sonora, Mexico, on the south from the Organ Pipe Cactus National Monument in Arizona on the north. The twelve-mile stretch of boundary from Monument 163 to 166 is readily accessible to grazing animals from south of the line and, for most of the period since 1893 has been open also to grazing on the north. Most of the vegetation, however, has little or no value as forage and this, combined with a general deficiency of water for livestock, has resulted in only light grazing use throughout this portion of the International Boundary.

Creosotebush occurs at all three sites and is dominant at Monuments 164 and 165, as it was in 1893. Only two other scrub species—foothill paloverde and ironwood—are growing today near Monument 164, with no grasses and only two ephemerals—plantain and Russian thistle.

The Monument 165 site supports a richer scrub flora than the area in the vicinity of Monument 164, with desert saltbush, triangle bursage, and tomatillo added to the creosotebush, paloverde, and ironwood of the previous site. Here, I also recorded a trace of a perennial grass—threeawn. Neither of these areas has shown any life-form change over the past ninety years.

Although the Monument 166 site, because of man and his "improvements," has changed markedly adjacent to the boundary line, I suspect that the general area may have changed little.

The soil here is moderately salty, a condition that restricts both floral variety and plant density. Although, as noted above, some creosotebush is growing here, it occurs much more sparsely than at the two previous monuments. Two salt-tolerant shrubs predominate—desert saltbush and threadleaf saltbush. In the median between the two roads that show in the recent picture, these two species, intermixed with three forbs—Russian thistle, fiddleneck, and plantain—provide an almost complete ground cover. Elsewhere, there are many areas of bare ground, but these seem to be less extensive than they were in 1893.

Although all three of these sites are probably somewhat better vegetated now than when the early pictures were taken, the differences seem to be slight and might be attributed to temporary precipitation deficiencies and excesses during the early 1890s and 1980s, respectively. In any event, there has been no evident change in life form during the intervening years.

MONUMENT NO. 164. Sonoran Desert

Monument No. 165. Sonoran Desert

MONUMENT NO. 166. Sonoran Desert

Monument No. 167. Sonoran Desert

Monument 167 was erected in what was then open desert, but today is the northwestern edge of the town of Sonoyta. Despite this urban encroachment and the fact that various buildings and a road lie less than a hundred yards east and south of the marker, the general area remains essentially unchanged from its appearance ninety years earlier.

Although no vegetation shows in the foreground of the original picture, the background contains what appears to be a species of cholla, as well as ocotillo, sahuaro cactus, an abundance of triangle bursage, and trees that are probably paloverde, ironwood, and velvet mesquite.

In my recent photo desert saltbush and buckhorn cholla are the only scrub species that show in the foreground. The background vegetation comprises largely triangle bursage, arborescent cholla cacti, sahuaro cactus, foothill paloverde, ironwood, and velvet mesquite.

Here, as at the previous three monuments, there has been no life-form change. Even differences in individual taxa are problematical and, at best, slight.

Monument No. 167. Sonoran Desert

Monument No. 168. Sonoran Desert

This masonry monument (No. 168) occupies the site of old No. 9, and stands prominently on a disjunct southern portion of the Sonoyta Mountains, one and a half miles west of Sonoyta.

The vegetation showing in the 1893 photo appears to be brittlebush, creosotebush, and one sahuaro. Other species may make up part of the plant cover but cannot be identified. Today's vegetation, which may in fact differ little from that of ninety years ago, consists largely of foothill paloverde (shown to the right and left of the monument), brittlebush, triangle bursage, buckhorn cholla, and creosotebush. Wedgeleaf limberbush, range ratany, Englemann hedgehog cactus, organpipe cactus, and sahuaro cactus also occur here, but less commonly.

Based solely on the evidence presented in the two pictures, one could conclude that paloverde has increased or even invaded this area, where none grew before. I consider even the first of these possibilities to be questionable, as this mountaintop is not readily accessible to either man or beast. Paloverde is a strong dominant of the Sonoran Desert on similar sites throughout this region, is rather long-lived and extremely drought resistant, and does not fluctuate widely in numbers over short periods of time. Thus, I would expect that other photographs taken nearby would have shown paloverdes, and, in fact, some of the larger plants that show on the skyline may be paloverdes.

In conclusion, I assume that little or no vegetational change has taken place here since 1893. Certainly, there has been no change in life form.

MONUMENT NO. 168. Sonoran Desert

Monument No. 169. Sonoran Desert

The next three monuments, 169–71, lie in La Abra Plain, in the Organ Pipe Cactus National Monument.

The two photographs show changes in details of the vegetation but no major modification in either floristics or life form. The area then, as it is today, was characterized by an overstory of ironwood, foothill paloverde, and sahuaro cactus, and an understory of triangle bursage and creosotebush. Each of the latter two species was more abundant when I took the repeat photo than any one of the three larger taxa. There were no grasses or forbs, either annual or perennial.

Monument No. 169. Sonoran Desert

Monument No. 170. Sonoran Desert

The Monument 170 site differs primarily from that at Monument 169 in its much lower plant density and somewhat greater number of common plant species. The same "trees"—ironwood, foothill paloverde, and sahuaro cactus—are scattered throughout here today, as they were formerly. The banks of a nearby wash also support a little ironwood, blue paloverde, and velvet mesquite. Creosotebush and triangle bursage dominate the lower-stature vegetation, as they did at the previous monument. In addition, however, there is a little jumping cholla (as in 1893) and desert saltbush. A large cholla can also be seen in the earlier photo; the saltbush may have been present but cannot be identified.

Monument No. 170. Sonoran Desert

Monument No. 171. Sonoran Desert

This site differs from the previous two in that it is on the well-drained slope of a small hill while the others were in essentially level and less rocky terrain. The consequent edaphic and soil-moisture differences may account for an abundance here of brittlebush and, perhaps, for the occurrence of organpipe cactus. In addition to these two, the vegetation consists primarily of creosotebush, triangle bursage, foothill paloverde, and an occasional sahuaro. These same taxa can be distinguished in the 1893 photo.

MONUMENT NO. 171. Sonoran Desert

Monument No. 172. Sonoran Desert

Monument 172 stands on a small prominence at the southern end of Quitobaquito Hills, about a half mile east of the well-known springs of the same name. As a spring of any size is a life-supporting rarity in the Sonoran Desert and as this is a multiple spring with a considerable and reliable, even though alkaline, flow, the Papago Indians, and later the Mexicans, have long used this area as a place of habitation, greatly preceding the time when the monument was erected.

The entire region near the spring has probably been grazed by domestic livestock since their introduction by the Spaniards in the early eighteenth century. Any grasses that might have grown there prior to that time had probably been grazed out long before the monument was erected.

Scrub vegetation characterized the area then, as it still does today. Even though the land north of the line now lies within the Organ Pipe Cactus National Monument, it, too, was grazed as recently as 1979. Because of the low palatability of most of the plants both north and south of the border, however, and the small amount of forage available, a rather good stand of the native vegetation still remains on both sides of the boundary fence.

When I visited the area in the spring of 1984, triangle bursage, white bursage, creosotebush, velvet mesquite, foothill paloverde, and ironwood predominated, with lesser amounts of sahuaro, jumping cholla, and organ-pipe cactus. These same species seem to have been present when the earlier picture was taken, thus no change in the general vegetation is indicated.

MONUMENT NO. 172. Sonoran Desert

Monument No. 173. Sonoran Desert

Monument 173, the westernmost on the Organ Pipe Cactus National Monument, stands on a small hill known locally as Cerro Blanco.

Three miles distant from Quitobaquito Springs or any other water, and thus subject to little or no pressure from domestic livestock, the vegetation here represents what might be expected as the climax for the area.

Triangle bursage, white bursage, creosotebush, and foothill paloverde predominate, with scattered individuals of sahuaro cactus, cane cholla, and ocotillo. Blue paloverde and ironwood grow on the banks of nearby washes. Although not all of these same species can be distinguished in the earlier photo, most of them can and, as they all are representative members of the climax for this ecosystem, probably were all present when the monument was erected.

Monument No. 173. Sonoran Desert

Monument No. 174. Sonoran Desert

This, the easternmost of the monuments on the Cabeza Prieta National Wildlife Refuge, is readily accessible from Mexican Federal Highway 15, which parallels the boundary here. The game refuge is protected from grazing by domestic livestock, as it has been for many years. This entire region on both sides of the boundary contains little or no water for livestock, a condition that contributes to the light or no grazing condition, both in the United States and Mexico.

When the earlier picture was taken there were few ephemerals, probably due to the preceding severe drought period. My repeat photo, in contrast, shows a good ground cover of this transitory kind of vegetation. Precipitation during the preceding rainy season had been heavier than normal, and the ephemerals had sprung up in response.

If there has been any change in perennial vegetation here during the intervening ninety years, the pictures do not so indicate. Creosotebush and (triangle?) bursage were the principal shrubs in 1893 as they are today. Ocotillo can also be identified in the earlier picture, as (probably) can foothill paloverde. Other shrubs that I recorded in 1984 as being common were wedgeleaf limberbush, jumping cholla, ocotillo, and ironwood. These also may have been there in 1893 but, except for the ocotillo, cannot be identified in the picture. The bursage growing there today is the so-called triangle variety and may be assumed to have been the one present in 1893.

Monument No. 174. Sonoran Desert

Monument No. 175. Sonoran Desert

Monument 175, another of the impressive masonry markers, was constructed on a hill at the site of former Monument No. 6.

Except for a better stand of ephemerals when I took the repeat photo, there has been no appreciable change in the vegetation in the vicinity of this monument since it was built. The principal species that I recorded at the site in 1984 were brittlebush and creosotebush. Less abundant, but still characteristic of the area, were buckhorn cholla (showing just left of the monument), teddybear cholla, wedgeleaf limberbush, ocotillo, and sahuaro cactus. The only grass was an ephemeral—six-week threeawn.

Monument No. 175. Sonoran Desert

Monument No. 176. Sonoran Desert

The vegetation at Monument 176, at the southern extremity of the Agua Dulce Mountains, seems to have changed little during the past ninety-one years. As with the two preceding markers, there were more ephemerals in 1984 than in 1893, but again, this represents a temporary response to precipitation rather than a real vegetational change.

The principal woody species that I recorded when taking the repeat picture were creosotebush, wedgeleaf limberbush (showing in left foreground), ocotillo, and foothill paloverde. Two perennial forbs were also common—silverbush and fagonia. The only grass was an ephemeral—six-week threeawn.

MONUMENT NO. 176. Sonoran Desert

Monument No. 177. Sonoran Desert

Monument 177 is located in the Davidson Canyon drainage, south and west of the Agua Dulce Mountains. As with the three preceding monuments, there has been no change in life form here since the monument was erected. And, again, there were more ephemerals at the time I rephotographed the monument than show in the original.

The principal species at this site in the spring of 1984 were creosotebush, white bursage, triangle bursage, ocotillo, ironwood, and foothill paloverde. There were lesser amounts of buckhorn cholla, wedgeleaf limberbush, and diamond cholla. Six-week threeawn was the only grass.

Monument No. 177. Sonoran Desert

Monument No. 178. Sonoran Desert

This monument is located on the north-facing slope of a granite ridge a short distance north of a massive basaltic outflow and volcanic crater dome.

There is no evidence of any vegetational change near the monument since it was erected. Today's vegetation (probably similar to that of 1893) consists primarily of creosotebush, foothill paloverde, ocotillo, triangle bursage, sahuaro, and rose mallow. When I visited the area in April 1984, there were no perennial grasses and only scattered plants of six-week threeawn.

Monument No. 178. Sonoran Desert

Monument No. 179. Sonoran Desert

Here, as throughout this area, no appreciable change in the vegetation during the period since 1893 was apparent when I examined it in 1984. The repeat photo, taken from a point closer to the monument than the original, consequently shows the hilltop vegetation in greater detail and does not include some of the foreground creosotebush and brittlebush (or bursage) visible in the original. Other woody species that show in the original are ocotillo, foothill paloverde, an arborescent cholla, and sahuaro. These same species occur there today, the arborescent cholla being buckhorn cholla. In addition, a single perennial forb—fagonia—is common. This may have been present earlier but would not have shown in the picture.

Monument No. 179. Sonoran Desert

Monument No. 180. Sonoran Desert

Monument 180 stands in an extensive flat that is strewn, or almost paved, with small basaltic rocks. The Sierra Pinta is the mountain range showing against the skyline in the background. The light-colored, almost lakelike area beyond the monument that shows most clearly in the original picture is a playa or shallow, usually dry, lake bed.

This basaltic plain is a poor habitat for plants, and the plant cover was, and is, sparse. The taxa are few and their density is low. The little vegetation that was here in April of 1984 consisted of creosotebush, foothill paloverde, ocotillo, and an occasional sahuaro. There were no perennial grasses on the flat plain, although scattered clumps of big tobosagrass characterized some of the nearby shallow drainages. Small areas of six-week threeawn grew here and there on the plain, some of which show in the recent photo.

Any vegetational changes that may have taken place here since 1893 are negligible and are not apparent from a study of the two pictures.

MONUMENT NO. 180. Sonoran Desert

Monument No. 181. Sonoran Desert

Four and 8/10 miles west of Monument 180, No. 181 stands on a low, rock-strewn hill in the Pinacate Lava Flow, about a half mile south of an extinct crater that is shown on U.S. Geological Survey Sierra Arida Quadrangle map as *Monument Bluff.*

The slight eminence where this boundary marker is located was sparsely vegetated in 1893, and still is today. What little vegetation there is consists of creosotebush, brittlebush, foothill paloverde, and ocotillo. Today's plants may be in a little better condition than when the earlier picture was taken; otherwise there has been no evident change in the vegetation.

Monument No. 181. Sonoran Desert

Monument No. 182. Sonoran Desert

Monument 182 is on the south-facing slope of a sparsely vegetated conical hill. A comparison of the two pictures shows an equally sparse stand of vegetation ninety-one years earlier.

Only two perennial taxa—creosotebush and ocotillo—occur here with any frequency today. These same two species can be distinguished in the earlier picture.

Monument No. 182. Sonoran Desert

Monument No. 183. Sonoran Desert

This monument was erected on a narrow, granite ridge, almost devoid of soil. The little moisture that falls here penetrates the fractured rocks rather well and has resulted in a more diverse flora than typified either the Monument 182 site or the Pinacata Lava Flow area.

Foothill paloverde, creosotebush, torote, brittlebush, white bursage, and ocotillo were most abundant in 1984, with buckhorn cholla and fagonia secondary. There was no ironwood, although it was common on the banks of washes throughout this area. There were no perennial grasses, but there was a little six-week threeawn.

A comparison of the two pictures seems to indicate more vegetation throughout this area today than in 1893. This, however, may reflect the quality and lack of contrast in the original, rather than any real vegetational differences. In any event, there has been no life-form change during the interim represented by the two photographs.

Monument No. 183. Sonoran Desert

Monument No. 184. Sonoran Desert

Monuments 184 through 186 are located high on rugged peaks or saddles in the arid Tule Mountains. They are difficult to reach, in part because of the remoteness of the area and the lack of roads, and in part due to the extreme heat and aridity of the area, the steepness and roughness of the mountains, and their considerable elevation above the more accessible lands below.

Monument 184 was placed in a high saddle 800 feet above the valleys to the north or south. As described in Senate Document 247 (p. 192):

> No. 184 . . . overlooks a vast extent of country eastward, the immediate neighborhood being a succession of rugged mountains, divided by deep, precipitous cañons. The steep, rocky sides of this mountain would not admit of even pack transportation, the pieces of the monument being carried by hand about 1¼ miles.

Because of the arid climate and excessively rocky character of the ground surface, plant density here is low. Study of the original picture suggests even fewer plants at the time the monument was erected. This impression, however, may be because of lesser detail decipherable in the earlier photograph. If there was, in fact, less vegetation then than now, this is the only apparent change in the plant cover.

When I visited the area in March 1985, the principal vegetation consisted of white bursage, creosotebush, brittlebush, trixis, ocotillo, and range ratany.

The only animal I saw was a solitary buzzard that had a nest in the nearby crags. En route to the monument I noted one set of bighorn sheep tracks.

Monument No. 184. Sonoran Desert

Monument No. 185. Sonoran Desert

Senate Document 247 says of this monument:

> On a still higher and more rugged mountain 2 3/4 miles beyond Monument 184 was located No. 185. The work here was still more difficult. Pack animals were used in carrying the sections of the monument part way up the slope and work completed by hand. This monument commands a magnificent view of rough and rugged mountain peaks rising in needle-like sharpness and separated by cañons whose sides are often vertical precipices.

Due to circumstances beyond my control—the manager of the Cabeza Prieta National Wildlife Refuge refusing to grant us permission to drive to the base of the mountain—I was unable to reach this monument. This was the only lack of cooperation we received from any individual or public agency during our traverse along the entire length of the boundary.

Fortunately, this did not adversely affect the study, as the original picture shows only rocks and a little disturbed soil around the base of the monument. The little vegetation that can be seen is on distant hills and cannot be identified. There is, however, no reason to assume any appreciable difference in the vegetation here from that at the adjacent monument sites either to the east or the west. As none of the adjacent sites show any vegetational changes to have occurred between 1893 and the present, there is little possibility here of change on a similar site.

Monument No. 185. Sonoran Desert

Monument No. 186. Sonoran Desert

This monument, as with No. 185, I was unable to reach. However, it was reached and photographed by a helpful friend, Pete Cowgill. The photograph of his that I have used was taken from the base of the monument, and, although it does not include the monument, it does give the background that was shown in the original picture.

Quoting again from Senate Document 247:

> No. 186 was placed where the boundary crossed the highest ridge of the Tule Mountains. It is 2 3/4 miles west of the preceding, and like the other its erection was attended with severe labor, requiring hard transportation a part of the way.

The terrain here is mostly rocks, and no plants are identifiable in the 1893 photo. Cowgill's data, recorded at the time of his April 1985 retake, listed as principal species essentially the same ones that had characterized the other monument sites in the Tule Mountains, namely: white bursage, creosotebush, foothill paloverde, ocotillo, and brittlebush. Other plants he saw growing in the general area were wedgeleaf limberbush, torote, an arborescent cholla, barrel cactus, and prickly pear.

Here, as previously, I have no reason to assume any vegetational change.

Monument No. 186. Sonoran Desert

Monument No. 187. Sonoran Desert

Monument 187, the westernmost of those in the Tule Mountains, was located on a disjunct basaltic portion of the range. The monument, although barely visible against the skyline to the naked eye from the camera station, cannot be made out in either of the photographs.

Erection of this monument was attended with many difficulties, as it was transported to the location entire, rather than in sections. The site was remote, and the terrain of malpais boulders provided a rough and hazardous route over which to transport the heavy and cumbersome iron obelisk.

I reached the site from Mexican Route No. 15 after walking miles across hills and washes before finally spotting the monument in the far distance. Much additional walking finally brought me to the original camera station. Perhaps Mr. Payne, the Boundary Commission photographer, was too exhausted to photograph the monument from close up; in any event, I was glad he took his picture from an adjacent small hill rather than from the top of the higher one.

The site is many miles from any human habitation, is totally devoid of water for livestock, and appears never to have been grazed by any type of domestic animals. The rather sparse vegetation consists almost entirely of brittlebush, white bursage, and creosotebush, with only an occasional foothill paloverde, ocotillo, and buckhorn cholla. The original photo indicates that these same species predominated when the monument was erected.

Monument No. 187. Sonoran Desert

Monument No. 188. Sonoran Desert

Monument 188 was placed in an extensive creosotebush flat in an area shown on the U.S. Geological Survey Tule Mountains Quadrangle map as the Lechuguilla Desert. This area, although not rendered inaccessible by mountainous terrain, is so remote from human habitation and so hostile environmentally that no boundary fence has ever been built east of this site as far as Monument 183 or west to a point a short distance east of Monument 192.

Creosotebush and white bursage predominate here as they did in 1893. Range ratany ranks third in abundance and may also have been present earlier but cannot be identified in the picture. In addition, foothill paloverde, ocotillo, and buckhorn cholla occur occasionally today.

Monument No. 188. Sonoran Desert

Monument No. 189. Sonoran Desert

This monument, located on the crest of Sierra de la Lechuguilla, eluded us. Additional hours of search might have revealed it, but I felt would not have provided any additional information on the vegetation. We reached a point suggested by the map as approximately that of the original camera station. From this location I took several pictures of the Lechuguilla Mountains lying to our west. One of these is shown as a substitute for a retake of the original.

The original picture was apparently taken from a point too far from the monument for it to be discernable in the photo. Most of the vegetation, however, appears to have been foothill paloverde and ironwood. These same trees occur generally throughout the area today. At the site of my repeat photo creosotebush, white bursage, ocotillo, and torote were also common and may be assumed to have been present but unidentifiable in the 1893 picture.

Monument No. 189. Sonoran Desert

MONUMENT NO. 190. Sonoran Desert

The lack of a boundary fence here made this monument a difficult one to find. After many hours of searching, however, I finally spotted it near the crest of a small hill nestled near the western base of the Lechuguilla Mountains.

The terrain here continues as rough, often inaccessible, craggy mountains with very little soil to support vegetation. As a consequence, the plant cover on the slopes continues as a sparse mixture of shrubs and trees in which only a few taxa constitute most of the plants that do manage to become established. When I photographed the area in the spring of 1985, six species—creosotebush, white bursage, ocotillo, foothill paloverde, ironwood, and torote—made up an estimated 99 percent of the plant cover. Although plant composition in the valley that shows on the left of the picture would differ in some respects from that on the slopes, I did not visit that area or take notes on its vegetation.

MONUMENT NO. 190. Sonoran Desert

Monument No. 191. Sonoran Desert

The lack of a boundary fence anywhere in this area, coupled with rugged terrain and lack of roads, all combined to make this monument difficult to locate. After two days of searching, I finally spotted it on top of one of the many peaks of the Tinajas Altas Mountains. Senate Document 247 (p. 193) says of this area:

> This range is one of the roughest and most precipitous encountered on the entire boundary. . . . The contrast between the valleys and steep mountains, whose sides are frequently vertical for many hundred feet, is more strongly marked than elsewhere along the boundary. The almost total absence of vegetation, sharp, rugged rocks only being visible, adds materially to the wild and desolate effect.
>
> No. 191 was placed 2 miles beyond No. 190, and marks the point where the boundary reaches its highest elevation in crossing these mountains. In the immediate vicinity are many other high ridges, with intervening chasms several hundred feet in depth. The summits of these ridges are so narrow that men found great difficulty in retaining their positions while at work on the survey. The monuments upon these rocky mountain tops were all of the sectional kind, and were bolted to the solid rock after the surface had been suitably leveled by blasting.

As with Monuments 187 and 189, the Boundary Commission photographer was satisfied with obtaining a distant, rather than a close-up, view of the monument. This saved him, and me, from making an arduous and hazardous 900-foot climb to the summit.

Ironwood, brittlebush, hind mariola, and chuparosa were abundant in the canyon bottom. Although the steeply sloping canyon walls supported little or no vegetation, the ironwood and brittlebush, accompanied by ocotillo and white bursage, tended to grow there also where pockets of soil or cracks in the rocks afforded a foothold. It can safely be assumed that these same species inhabited the area at the time the monument was installed.

Monument No. 191. Sonoran Desert

Monument No. 192. Sonoran Desert

Here, also, as with some of the preceding monuments, Mr. Payne, the Boundary Commission photographer, was content to take his photograph from a distance. The monument, which cannot be distinguished in either of the pictures, was located near the western edge of the Tinajas Altas Mountains.

The vegetation here continues much as it has in the mountains of this region to the east. The principal species that were present when I took the repeat photo in the spring of 1985 were creosotebush, white bursage, foothill paloverde, brittlebush, teddybear cholla, sahuaro cactus, and ironwood. Most of these same species seem to have been present when the original picture was taken.

MONUMENT NO. 192. Sonoran Desert

Monument Nos. 193–204. Sonoran Desert

These twelve monuments all lie in an extensive plain that is shown on the U.S. Geological Survey maps as the Yuma Desert. Shreve includes this area as a portion of the Lower Colorado Valley, a subdivision of the Sonoran Desert (Shreve, 1951). The gradient over the twenty-eight miles of boundary represented here slopes gently to the west and the Colorado River. Because of the low precipitation, sandy soil, and gentle gradient, however, little runoff ever reaches the river.

Although only a single species, creosotebush, occurs at each of the first eleven of these monuments, there is sufficient taxonomic similarity to permit their being handled as a unit. Thus, white bursage occurred at ten of the twelve and California threeawn at nine. Big tobosagrass, a coarse perennial that is generally restricted to sand dunes and the banks of drainages in this area, was recorded at five of the monuments and might have been found at most, or all, of them with more diligent search. Palmer coldenia, a perennial forb, also was a characteristic species at five of the sites and might have been recorded at others with more extensive search. Other taxa that occurred occasionally were ocotillo and brittlebush (two monuments each), and teddybear cholla and range ratany (one monument each).

The arid and hostile environment, as well as the uniformity of habitat, are indicated both by the few perennial species encountered and by the low density of all perennial taxa throughout the area. As Shreve observes, ephemerals are abundant in the Lower Colorado Valley in season. Their prevalence, however, is highly dependent on seasonal precipitation. Most of the ground cover showing in some of my 1984 photos, 195–97 for example, consists of ephemerals that I did not attempt to identify.

There seem to have been no essential vegetational or erosional changes throughout this area during the past ninety-one years. Many differences in individual plants have occurred, as a comparison of the 1893 and 1984 photos of the various monuments will indicate. Note, also, at Monuments 195 and 197, the paucity of ephemerals in the original as compared with their relative abundance in the retakes.

When the monuments were erected, Number 203 was in open, creosotebush desert; today it stands on the eastern edge of the town of San Luis. The native vegetation remains largely unchanged, but signs of urban development are almost everywhere evident.

Monument 204 was built in open desert; today downtown San Luis surrounds it. Buildings and paved streets now adjoin the monument; graffiti adorns it, and the original creosotebush has been replaced by exotics—fan palms, oleander, and casuarina.

MONUMENT NO. 193. Sonoran Desert

MONUMENT NO. 194. Sonoran Desert

MONUMENT NO. 195. Sonoran Desert

Monument No. 196. Sonoran Desert

Monument No. 197. Sonoran Desert

MONUMENT NO. 198. Sonoran Desert

MONUMENT NO. 199. Sonoran Desert

Monument No. 200. Sonoran Desert

MONUMENT NO. 201. Sonoran Desert

Monument No. 202. Sonoran Desert

Monument No. 203. Sonoran Desert

MONUMENT NO. 204. Sonoran Desert

Monument No. 205. Riparian (originally)

U.S. Customs at San Luis told us that Monument 205 is no longer there. I searched the area indicated 1.86 miles west of Monument 204 and could find no trace of it, so I assume it has vanished.

The original picture shows a dense, riverbottom stand of cottonwoods, or cottonwoods and willows. Today, open fields, an irrigation canal, roads, and small buildings occupy the area.

Monument No. 205. Riparian (originally)

Discussion

Research such as this study embodies does not lend itself to ready and conventional analysis of the sort involved where various treatments are applied and the results compared with those from untreated areas. Such methods permit mathematically expressed conclusions and a calculated range of experimental error.

In the present instance, where there were no treatments, where no exact measurements were made, and where photographs taken over a span of many years constituted the main data source, experience and familiarity with the region and its ecological characteristics become essential factors in reaching sound conclusions. The experience and judgment of the researcher are of particular importance in studies of these sorts where changes are being related to causes.

Each pair of photographs analyzed here shows vegetational changes. Most of these are in the nature of the death and disappearance of individual plants and the establishment and growth of others. In some instances, however, there has been a change in density of a particular life form or a partial or complete replacement of one life form by another. Or, there may have been evident erosional activity during the time period involved.

Relating these changes to specific causes becomes somewhat subjective even with the most sincere attempt to be objective. Conclusion validity will be affected both by the background of the research worker and by the thoroughness with which he reviews the pertinent literature. In recognition of these facts I proceed with hesitation to suggest possible reasons for the vegetational changes recorded.

Any one or more of four causes may have been largely responsible for the changes observed: (1) a change in climate, (2) grazing, (3) fire, and (4) urban or rural development. The order in which I have listed these does not indicate their possible relative importance.

Discussion

CLIMATE

The effect of climate on a change in the perennial vegetation over an essentially short ninety-year period is difficult to determine with any degree of certainty. The vegetation of the arid regions under consideration here has evolved under, and become adjusted to, aridity and recurrent droughts. A few years of subnormal rainfall or even a severe drought adversely affect the dominant species and may kill some of them. They do not, however, typically die out to be replaced by another life form. More commonly, they will be replaced by others of the same species or, at least, of the same life form. Even the ephemerals that disappear, sometimes for several years, appear in abundance when the rains return.

Grasses are more likely to be adversely affected by a severe drought than are most of the woody desert plants with their deeper root systems. Where grassland areas are surrounded by or adjoin aggressive scrub communities a drought severe enough to kill most or all the grasses would probably benefit the competitive shrubs. This would be particularly true in the absence of occasional fires that typically harm woody species more than grasses, or where excessive numbers of livestock selectively overgraze the grasses.

The moisture available to plants is affected not only by the amount of precipitation but also by temperature and wind velocity. Thus even if precipitation amounts should remain constant, a marked increase in temperatures and/or wind velocities would result in greater evaporation and in less moisture being available to the vegetation. The reverse also holds.

Climatic records, whether summarized as annual, seasonal, monthly, or even daily precipitation or temperature parameters, often do not indicate moisture available to the vegetation. Thus, precipitation that shows as one inch falling on a given day may all have fallen over a period of up to twenty-four hours and largely been absorbed by the soil. Or, rain falling on already saturated soil, on steep slopes, or on impermeable soils may largely be lost. Climatic records have their limitations; they show meteorologic data but not necessarily these data as experienced by the plant. Any attempt to relate climate to vegetational changes must do so with these facts in mind.

There has been a gradual, long-continued trend toward greater aridity in the North American Southwest. This began at least as long ago as late Wisconsinan time—22,000 to 11,000 years B.P. (before the present by radio-carbon dating) (Van Devender and Spaulding, 1979). Climatic and vegetational changes resulting from recession of the late Wisconsinan ice sheet continued into the late Holocene about 8,000 B.P. At about that time "the present climatic and vegetational regions were established. . . . Winter precipitation was reduced in or withdrawn from much of the Southwest. The summer monsoon expanded, resulting in the present geographic difference in seasonality of rainfall and the related segregation of the biota" (Van Devender and Spaulding, 1979, p. 708). This gradual increase in aridity and change in precipitation seasonality during the Holocene had a definite prehistoric effect on the vegetation.

This long-term, slight tendency toward greater aridity seems to be continuing into the present. In an analysis of eighteen Arizona and New Mexico stations, Sellers (1960) concluded: "Since 1921–24 the 20-year average annual precipitation in Arizona and Western New Mexico has decreased by about 25 percent. . . ." The period of below-average precipitation that prevailed in 1959 when Sellers's study was concluded, continued through 1965 (Betancourt and Turner, 1985). Cooke and Reeves (1976) review Sellers's results, together with temperature data from Hastings and Turner (1965), concluding:

> Since the turn of the century there has been a slight decrease in summer precipitation and a greater decrease in winter precipitation, together with general slight increases of temperature. These trends point towards greater aridity since 1898, reflected perhaps in greater evaporation and less soil moisture.

These decreases in precipitation have been too slight to be statistically significant as indicated on page 78 of Cooke and Reeves:

> There have been no statistically significant secular changes in annual, summer, or annual non-summer precipitation totals during the last hundred years at the stations studied in southern Arizona.

Most southwestern grasses respond primarily to summer rather than to winter rainfall. As a consequence, forage production from perennial grasses depends largely on the quantity and quality of summer precipitation (Culley, 1943; Reynolds, 1954). Forage production involves seed production and the development of a root system that will be adequate for future needs. Summer droughts, therefore, would seem to be particularly inimical to the maintenance or establishment of most southwestern grass species.

In an analysis of summer precipitation (June–September) at the Ft. Lowell-Tucson stations, Cooke and Reeves (1976) show thirteen periods of summer precipitation below the mean between 1884 and 1960. Two of these, 1884–86 and 1899–1906, were particularly severe, with cumulative deficiencies of 8.834 and 10.794 inches, respectively. By comparison, the relatively short-term but severe drought of 1891–92 that occurred just prior to the erection of the monuments and that resulted in death by starvation of thousands of cattle in southern Arizona and New Mexico had a cumulative deficiency of only 3.016 inches.

What then, can we assume with regard to a recent change in climate (1885 to 1985) as being responsible for the vegetational changes observed here in at least the Chihuahuan Desert, the Evergreen Woodland, the Semidesert Grassland, and the eastern portion of the Sonoran Desert? There has been a trend away from grassland toward scrub, and the increasing aridity, including periodic severe summer droughts, may be in part responsible for this change.

Grazing

Cattle had been introduced into the Southwest long before the present boundary location was agreed upon, or the monuments that now mark it were established. The first domestic livestock were brought into the area by Francisco Vasquez de Coronado in 1540 (Wagoner, 1952). Coronado, with a retinue of at least 230 horsemen and 62 armed infantrymen, was seeking the fabled Seven Cities of Cibola. When he set out from the Pacific coastal city of Compostela on February 23, 1540, he had with him as a source of food en route 5,000 sheep and 150 head of hardy Andalusian cattle. In addition, there would have been at least 230 horses for his cavalry plus a remuda of replacements.

Although many of these cattle were abandoned between Culiacán and the Sinaloa River and multiplied there to form large herds in later years, Wagoner questions whether few, if any, cattle remained with Coronado by the time he reached southern Arizona and the Gila River.

The first major introduction of cattle, as well as of horses, sheep, and goats, into what is now southern Arizona and northern Mexico was made by Padre Eusebio Francisco Kino in the latter years of the seventeenth century (Bolton, 1932). At his initial Pimeria Alta mission of Nuestra Señora de los Dolores (the Dolores Mission) on the San Miguel River near the small Indian village of Cosari, he raised large numbers of livestock and provided the animals for missions and ranches at at least eighteen other locations in the area. These were distributed widely over present northern Sonora, Mexico, and southern Arizona, at "Caborca, Tubutama, San Ignacio, Imuris, Magdalena, Quíburi, Tumacácori, Cocóspora, San Xavier del Bac, Bacoancos, Guebavi, Síboda, Búsanic, Sonóita, San Lázaro, Sáric, Santa Bárbara and Santa Eulalia" (Bolton, 1932, p. 65).

Padre Kino had no livestock of his own; these were mission animals that he distributed to promote the establishment of other missions or to serve as a source of food, transportation, or textiles to others.

The large numbers of livestock distributed by Padre Kino is indicated in a letter written in the spring of 1702 to Father Visitor Antonio Leal (Bolton, 1948, Pt. 1, p. 357):

> There are already many cattle, sheep and goats, and horses; for, although in the past year I have given more than seven hundred cattle to the four fathers who entered this Pimeria, I have for the other new conversions and missions . . . more than three thousand five hundred

more cattle; and some of them are already far inland, ninety leagues from here, and by the divine grace they can pass with ease to the Californias, Upper and Lower. . . .

Padre Kino's livestock distribution efforts all seem to have been restricted to areas south of the Gila River, in Pimeria Alta. In that region however, they were widespread and undoubtedly formed at least one nucleus for the large numbers of animals that later travelers encountered along what is now the border between the United States and Mexico.

Following the early Jesuit era gold and silver were discovered in Pimeria Alta. These discoveries resulted in the establishment of several large haciendas during the eighteenth and nineteenth centuries in the productive valleys of the area (Haskett, 1935). The haciendas were stocked with cattle and other livestock that flourished on the generally abundant forage. One of the largest of these, the San Bernardino, located on the present International Boundary about seventeen miles east of Douglas, and the greater part of which still lies in Mexico, is said in its heyday to have grazed 100,000 cattle, 10,000 horses, and 5,000 mules (Haskett, 1935, p. 6). Another, the Babocómari, in a drainage of the same name that leads into the San Pedro River, was reputed to have had 40,000 cattle in addition to an unknown but large number of horses and mules.

During the early nineteenth century while international proprietorship of the area was very much in a state of flux, Apache Indian raids became so frequent that these and most other land grants in Pimeria Alta were abandoned and the cattle allowed to run wild. The largely Andalusian cattle readily adapted to the wild habitat, the bulls in particular proving extremely dangerous to later travelers in the region.

By the time of Captain Philip St. George Cooke's march with the Mormon Battalion in 1846 and 1847, despite the hazards posed by wild bulls on the San Bernardino Ranch and elsewhere, the Apaches had probably greatly reduced the numbers of cattle. Cows and young animals were more easily killed than the bulls, and this may account for the large concentrations of bulls reported by Cooke and others.

During the mid-nineteenth century, few animals remained from the original large herds. These were augmented by small herds that were driven into the area or were en route to California. Despite these sources, however, cattle numbers were in general low. Apache raids, combined with forays by Mexican and other cattle rustlers, made life throughout the area outside walled presidios or pueblos hazardous and grazing of livestock on the open range impossible. Not until General George Crook's campaign of 1870–72 against the Apaches were they finally subdued and the gate opened to general use of the open range by domestic livestock.

Cattle numbers increased rapidly during the next few years until most of southern Arizona's rangelands were heavily overstocked even prior to the severe drought of 1891–92. In addition to the unrecorded thousands of head that died from starvation during that drought, an estimated 200,000 head of cattle were shipped from Arizona during the twelve months ending with June 30, 1893 (Wagoner, 1952, p. 53).

Although the numbers of cattle presumably grazing southern Arizona's ranges are available from the county assessor's records, these must be viewed with skepticism. They show the cattle reported for tax purposes and, as there was little or no way of checking on unreported animals, probably represented a fraction of the actual numbers. As an example, two figures for Pima, Santa Cruz, and Cochise counties in southern Arizona are available for 1900—the first the counties' tax-assessor figure, the second that of the U.S. Census Bureau. The county figure gives 191,190 head, the census 316,813, suggesting that for tax purposes at least, 125,623 animals were not reported. This raises the question: How many were also not reported to the U.S. Census taker?

Haskett notes (1935, p. 41), with reference to the reliability of the tax assessor's figures:

> Cattle production in Arizona seems to have reached its peak in 1891. The total number on the assessment rolls for that year was 720,941 head. . . . Taking one thing with another, however, it was the opinion of men who knew the facts in the case that there were fully 1,500,000 cattle on ranges in Arizona that year.

These large numbers, combined with one of the worst droughts on record, proved ruinous to the cattle industry in southern Arizona, cattle losses being estimated at from 50 to 75 percent of southern Arizona's range cattle (p. 42).

Sheep, rather than cattle, seem to have been the principal kind of livestock during the early years in New Mexico. Arizona had some sheep also, but not in the numbers or the proportion in relation to cattle, horses, and mules as New Mexico. In his discussion of livestock numbers in relation to gullying in New Mexico, Denevan (1967) devotes his livestock discussion almost entirely to sheep, making only one footnote mention of cattle. As he points out, sheep were introduced into New Mexico by Juan de Oñate in 1598, the increase spreading from the missions to the ranches. The numbers grew until by 1880 there were nearly 4 million sheep in the state. Citing U.S. Census figures for that year, Denevan contrasts this large number with a total of approximately 348,000 cattle. No figures were available on horses and mules.

Most of the sheep were grazed in the central and northern parts of the state, and, although there must have been some sheep as well as cattle in the south, particularly in the Palomas and Animas Valley areas, I have no record of these. There would seem to have been no sheep in either extreme southern New Mexico or Arizona in 1892 and 1893 or some mention would presumably have been made of them in Senate Document 247. The document (p. 183) does mention cattle grazing in Palomas Valley as an industry "badly crippled by the late droughts." In describing the Mosquito Springs area near the south end of the meridian section of the boundary, the document also says that these springs provided water for "large herds of cattle and horses" (p. 184). The region from here westward through Animas Valley to the Guadalupe Mountains at the Arizona–New Mexico border is generally open and accessible to grazing animals and would have been grazed by cattle. It also contained numerous herds of deer and antelope.

It would seem that the entire boundary region, with the exception of the far western area, has been subjected to varying degrees of grazing pressure by domestic livestock since Kino's introductions of the seventeenth century. This is particularly true of the period from about 1870–80 to the present. In general the more forage that has been available, the greater the availability of water, and the less rugged the terrain, the greater has been the grazing pressure.

Livestock numbers on both sides of the border have varied through the years and from place to place. For the most part, however, grazing pressures in some degree have persisted for over a hundred years. This long-continued preferential use of grasses as a life form over woody plants cannot help but have had a differential effect on the two life forms. Cattle graze most grasses in preference to most woody plants. When the dominant scrub species of the Chihuahuan Desert tend to be such entirely unpalatable species as creosotebush, mariola, and tarbush or such low-palatable, durable, and aggressive species as honey mesquite and spring whitethorn, long-continued inexorable differential grazing pressures will, over time, operate in favor of the scrub. The scrub species are favored here further by their overall abundance and a climate favorable to their establishment and growth. Under an environment where either grasses or scrub are able to grow, any pressure that favors one over the other will tip the scales in favor of the less pressured form. This seems to have occurred in the Chihuahuan Desert at those many monument locations where woody species have either entirely replaced grasses or where former grass-scrub communities have lost the grasses and now grow only scrub. Although the basic data of this study do not extend to a broader generalization, it may well be that additional extensive portions of this desert that today contain few or no perennial grasses, were largely grassland a hundred and more years ago.

The same principle applies in particular to the semidesert grassland. Here, too, is a zone of contention where either grasses or woody species are capable of establishment and growth, an area where grazing and other pressures, even though moderate, will, if continued long, favor one life form over the other and will in time change the general vegetational aspect from grass to scrub. This has happened over thousands of acres in the

Southwest, extending from Texas westward through southern New Mexico and Arizona and into Sonora, Mexico (Humphrey, 1958).

Grazing pressure may not be the only factor responsible for all of the scrub invasion in the semidesert grassland, but it does seem to be in large part responsible for the change. And, as in the Chihuahuan Desert, there were woody species around the edges and in permeating drainages waiting and ready to invade when conditions permitted. Chief among these in Arizona was velvet mesquite whose seeds were widely distributed in the droppings of grazing cattle, and that has transformed extensive areas of former grassland into a savanna type of woodland vegetation.

The Sonoran Desert today is characterized by various cacti, shrubs, and low-growing trees with comparatively few grasses. This study, however, reveals that there have been drastic changes in the ratio of grasses to woody plants during the last ninety years over approximately the eastern 40 percent of the Sonoran Desert portion of the boundary. Over the western 60 percent there have been few or no life-form changes.

When the monuments were established in 1892 and 1893 the areas surrounding most of those in the eastern portion of the Sonoran Desert (excluding a few located in extremely rugged terrain largely inaccessible to domestic livestock) grew a great deal more grass than they did ninety years later. In some instances, as at Monuments 143 and 144, what was then open grassland now grows no perennials except velvet mesquite and other woody vegetation. Even in 1956, thirty-six years earlier, the vegetation in that area appeared much as it does today (Humphrey, 1974, p. 391).

Analysis of land jurisdiction along this portion of the boundary reveals that the eastern 40 percent lies within the boundaries of the Tohono O'odham Indian Reservation, while the western 60 percent crosses the Organ Pipe Cactus National Monument, the Cabeza Prieta National Wildlife Refuge, and the Luke Air Force Range. The Tohono O'odham Reservation has been grazed, sometimes heavily, ever since cattle were first introduced into the area. Until recently the national monument has had some grazing but receives none now. The game refuge and the gunnery range are likewise ungrazed. Except in the immediate vicinity of Sonoyta and the nearby Quitobaquito Springs there is little or no water available for livestock, a factor that has always largely precluded the possibility of grazing livestock throughout most of this extensive area. The extreme aridity and sparseness of palatable forage species have also tended to prevent livestock use.

Thus, most of the western 60 percent of this portion of the boundary has received little or no grazing by domestic livestock, while the Tohono O'odham 40 percent has been grazed to and beyond its sustained-yield capacity for a long period.

As a pertinent item here, I directed a range survey (analysis of available forage and range condition) of the Tohono O'odham Reservation in 1936 and 1937. At that time we found the reservation to be overstocked, and extensive areas were so poorly suited to use by domestic livestock that more than 640 acres (one square mile) were required to provide enough feed to sustain one cow for a year.

The gradual increase in aridity noted earlier as having a possible effect on a change from grass to scrub in the Chihuahuan Desert and the semidesert grassland may also be in part responsible where similar changes have occurred in the Sonoran Desert. However, the grazing history on the Tohono O'odham suggests long-continued grazing as probably the most important factor here in the almost complete annihilation of grasses from areas where they were once dominant or where they occurred abundantly.

The same gradual increase in aridity with time may have characterized the western 60 percent of this portion of the boundary. This, of itself, however, has not been enough to cause a change in either life form or dominant taxa. The two xerophytic grasses, big galleta and California threeawn, that are locally characteristic, persist despite the vicissitudes of climate.

Fire

Natural (usually the result of lightning) and man-caused fires have affected the earth's vegetation for as long as there has been enough fuel to carry fire. Its importance in relation to the vegetation the world over varies from none to extreme depending on many variables, a discussion of which lies outside the purview of this study. In those parts of the American Southwest where fuel was plentiful and summer lightning storms frequent, it has been assumed, and to some extent documented, that natural fires were a frequent occurrence. Prior to the arrival of modern man these would have continued to burn until extinguished by rain or lack of fuel (Bahre, 1985).

Natural fires were augmented to some extent (just how much is not known) by Indians who used fire in various ways, as an aid to hunting, just for the spectacle, or, occasionally, in conjunction with warfare.

The frequency with which these early burns occurred and their effect on the vegetation have long been topics of study, discussion, and, often, of disagreement (Sauer, 1950; Stewart, 1954, 1955; Humphrey, 1958; Hastings and Turner, 1965; Cable, 1967; Martin, 1973).

There is general agreement that in an area capable of growing either grasses or low-growing trees and shrubs, fire not only favors the grasses but may keep the woody plants under control or entirely prevent their invasion. It has long been my conviction that this kind of control was highly effective in extensive portions of the semidesert grassland prior to the introduction of domestic livestock (Humphrey, 1958).

Some who question that fires occurred often enough to be this effective make the point that were this the case, early reports, diaries, and newspaper accounts would have contained more references to the grassland fires. In defense, I submit that usually only the large, spectacular fires were mentioned, as might be expected. The commonplace, whatever its nature, receives little attention. And, in a time when the region was largely or entirely unsettled, there were few or no scribes to observe, much less to record, the commonplace or, often, even the unusual.

Bahre (1985), on the other hand, in a thorough perusal of newspaper accounts of wildfires in southeastern Arizona between 1859 and 1890, recorded thirty-six separate writeups. In one instance three articles referred to the same fire and in another, two pertained to a single fire. Even with this slight duplication, however, thirty-three separate fires were sufficiently large to merit space in newspapers from Tubac, Tucson, Globe, or Tombstone. This averages a reportage of slightly more than one per year of fires large or spectacular enough to merit newspaper space. One cannot help but wonder how many others that were smaller or in remote areas far removed from population centers went unreported. The number may be expected to have been considerably in excess of those that were reported.

How effective these early fires may have been in controlling the shrubs in the Chihuahuan Desert is a moot question. I hazard the opinion that they may once have played a part locally but probably have been of little or no importance since the severe overgrazing that began during the latter part of the nineteenth century. Overgrazing and increased aridity would seem to have had a much greater effect.

In the semidesert grassland, on the other hand, fires probably originally played a major role in maintaining the grasses. Ever since the area became settled and stocked with domestic livestock in the 1870s and 1880s, however, they have probably become less and less of a factor. In many areas close grazing reduced the potential fuel; in others, fires would have been extinguished because they were seen as consuming valuable forage. Finally, with the birth of the U.S. Forest Service in 1905, fire became a dirty four-letter word, not only in the higher timbered areas, but in the adjacent grasslands. Since then, extensive fires have occurred only occasionally in these grasslands and have long ceased to be an important factor in the control of mesquite or other scrub species.

Fires burned unchecked in the evergreen woodland prior to the latter part of the nineteenth century. Fuel, in the form of trees, shrubs, and grass, is abundant, and lightning strikes are frequent in these high elevation areas (Arnold, 1963; Biswell, 1974, pp. 321–64). With the development of a fire-control philosophy about the turn of the twentieth century, free-running fires would have been suppressed, as they are today. This suppression has benefitted the trees and shrubs, permitting them to increase in both stature and density.

Fire also may have played some role in maintaining the once rather extensive Sonoran Desert grassland areas on the Tohono O'odham Indian

Reservation that now support scrub species with few or no perennial grasses (Humphrey, 1974, pp. 380–83). As has been indicated previously, however, grazing probably was more important than fire in effecting these changes.

Urban or Rural Development

The effects of urbanization on the vegetation are typically drastic and quite evident. The pressure from nearby population increases, often augmented by local livestock, such as horses, burros, goats, or milk cows, may entirely eradicate the grasses, if any, that formerly prevailed. The denuded ground is thus left open to invasion by such so-called weeds as Russian thistle, mustard, snakeweed, and burroweed. Other, even more drastic, changes include clearing the original vegetation for road rights-of-way, airplane landing strips or house construction. Urban development of any sort almost always results in an end to the former vegetation.

Rural developments, in contrast to urban, are often beneficial, at least from a point of view of ground cover, forage production, and erosion control. Three kinds of rural developments were observed along the boundary: plowing of the native vegetation preparatory to reseeding to a crop such as cotton, killing of the woody plants, and reseeding to exotic grasses.

In only two instances had the land been plowed, one near the south end of the meridian line, the other a few miles west of Columbus, New Mexico. Both were in heavy-soil, essentially level areas that had formerly grown tobosagrass. As a consequence, despite the bare soil, neither had suffered from accelerated erosion.

Semidesert grassland areas that have been invaded by mesquite or other woody plants may have the trees and shrubs killed by spraying with toxic chemicals or by uprooting, using bulldozers or the technique known as chaining. One instance of mesquite removal was encountered on the Bel Aire ranch east of Sasabe, Arizona. This range had then been reseeded to exotic lovegrasses, a combination of treatments that greatly increased the ground cover of grasses.

A second area, immediately east of the Huachuca Mountains, had been reseeded to lovegrasses without removing the woody overstory. Despite continued competition from the mesquite the reseeding here also resulted in a marked increase in grass ground cover.

Summary

Photographs taken in 1892 and 1893 of the United States–Mexico International Boundary monuments between the Rio Grande and Colorado River were compared with repeat photos taken ninety years later. The objective: to determine vegetation and landform changes that might have occurred during the interim.

As each repeat photo was taken the dominant vegetation at each site was recorded. The objective: to provide an accurate record of today's vegetation for use in this study and to serve as a 1983–84 historical record for future similar comparative studies.

As a supplement to the early photographic record, historical and scientific publications were reviewed and analyzed. Recent scientific publications were also reviewed to provide a more adequate basis for determining the possible causes of any changes observed.

Changes are particularly evident in the Chihuahuan Desert, the semidesert grassland, and the Sonoran Desert. In the Chihuahuan Desert many areas that supported grasses or a grass-scrub mixture now support only scrub. No areas that once were scrub now grow grasses. This one-directional movement appears to have resulted from a slight but consistent climatic trend toward increased aridity combined with long-exerted grazing pressures.

In the semidesert grassland, invasion of former open grassland by mesquite and other woody plants long has, and still is, converting many of these areas into a savanna type with low-growing trees and a variable grass understory. This change is attributed to grazing by domestic livestock and, in the earlier period, to a reduction in range fires.

Some of the evergreen woodland areas support a much denser stand of trees and shrubs today than formerly. This is particularly true at the higher elevations; the lower elevation areas seem, in general, to have changed little or not at all. No differences over time in life form were observed in this life zone.

Three factors are suggested as possibly contributing to the changes in the evergreen woodland: (1) areas immediately adjacent to the monuments would have been cleared to facilitate their construction; (2) with the passage of time the vegetation has normally grown thicker and taller; (3) fire control during much of the present century has benefitted the trees and shrubs.

In the eastern 40 percent of the Sonoran Desert portion of the boundary there has been a general and marked change in life form. The perennial grasses or grasses and scrub that once prevailed over much of this area have in large part been replaced by pure, or nearly pure, scrub.

There have been no life-form or appreciable taxonomic changes along the largely ungrazed 60 percent of the boundary west of the Tohono O'odham Reservation. Any possible increase in aridity with time has been too slight of itself to have appreciably changed the vegetation.

As the Sonoran Desert life-form changes were restricted to the Tohono O'odham Indian Reservation where grazing has long been practiced, domestic livestock use is believed to be the major cause of these changes.

Erosion since 1893 has been most active in the dune area west of the Rio Grande, most of the first fifty miles showing more soil movement than when the monuments were established. By comparison there were few signs of increases in erosion by water. Even the channels of the two major drainages, the San Pedro and Santa Cruz rivers, despite the apparent washing out of one monument (No. 118), appear to have changed little during the past ninety years.

Acknowledgments

During the course of the fieldwork we received excellent cooperation and assistance from various agencies and individuals. Initially, the entire study was made possible by grants from the National Geographic Society that covered both field and office expenses. We are deeply grateful for this financial assistance.

We want to express our appreciation to various federal officials and the agencies they represented and without whose cooperation the study could not have been completed. Representatives of five of these agencies cooperated in providing land-entry permits or, in one instance, in providing equipment and personnel that enabled us to reach an otherwise almost inaccessible area. Because we were carrying on the fieldwork along the International Boundary at a time of extensive illegal alien entry and considerable drug-running, we advised the U.S. Border Patrol whenever possible of our anticipated daily areas of operation. In addition, the first fourteen monuments west of El Paso lay primarily in a region of mobile sand dunes that made the area inaccessible to our VW bus. Here the El Paso Border Patrol office offered to provide a driver and one of their four-wheel-drive vehicles for the period required for me to photograph these monuments and record the necessary data on vegetation and soil movement. This assistance was essential here and the help given by Border Patrolman Joe F. Brewster was greatly appreciated. A second border patrolman, Kent E. Kooi, from Lordsburg, New Mexico, also went out of his way in providing helpful information.

On the Tohono O'odham Indian Reservation, Gu Vo District Chairman William Lewis Sr. and Chukut District Chairman Ramon Chavez granted free access to their districts in carrying out the study.

Harold J. Smith, superintendant of the Organ Pipe Cactus National Monument, and his personnel were highly cooperative in putting the monument and its facilities at our disposal.

The U.S. Marine Airbase at Yuma, Arizona, cooperated by granting access to the Luke Airforce Gunnery Range southeast of Yuma.

And finally, permission was granted by the Cabeza Prieta National Wildlife Refuge to enter that area in furtherance of the study.

I would like to thank three ranchers and their families in particular for their genuine welcome and support of the study. Mary and John Magoffin, at their ranch east of Douglas, Arizona, helped us get off to an optimistic start by their ready welcome and help in reaching three of the monuments. Dr. and Mrs. Lyle Robinson, owners of the Tres Bellotas Ranch in Arizona, not only welcomed us to their ranch but Lyle provided horses and rode with me the better part of two days to reach some of the less accessible monuments. Cordy and Bill Cowan and their daughter at the Cloverdale Ranch in southwestern New Mexico gave permission to camp at a choice site on their ranch and helped us locate some of the monuments.

Lonnie Moore, manager of the extensive Gray Ranch in southwestern New Mexico, also gave permission to travel over the Gray Ranch, an area closed to the public. Steve Dobrott, wildlife technician on the Gray Ranch, gave generously of both his time and knowledge of the country on two occasions during our sojourn in that remote corner of New Mexico. Mahlon T. Everhard, owner of the Hatchet Ranch, south of Hachita, New Mexico, was helpful in pointing out passable trails ("roads") to some of the so-called Meridian Boundary monuments. To all these hardy individuals we want to express out very real gratitude and thanks for the support and help they rendered.

We also want to thank Tad and Mary Jane Nichols for a weekend they spent with us in their jeep covering some of the (to our VW bus) inaccessible terrain in southwestern Arizona. Thanks, also, to Pete Cowgill, nature writer for the Arizona Daily Star, for sharing another weekend with us in similar terrain and for helping to acquaint the public with some aspects of our project; and to my former co-worker Bob Wagle and his wife Karen for their companionship and help on the Luke Air Force Gunnery Range.

My thanks also to Ray Turner, Julio Betancourt, and Tony Burgess, who went that extra mile in providing meteorologic data.

I am indebted to Dr. Charles Mason, Herbarium Curator, for his neverfailing answers in identifying "unknowns" from my ecological type of plant specimens.

Finally, my very real appreciation of the enthusiastic reception and painstaking editing given this manuscript by Dana Asbury and the University of New Mexico Press.

Appendix A

Distance between monuments

Monument No.	Distance Meters	Distance Miles	Monument No.	Distance Meters	Distance Miles
1	—	—	24	4,436	2.76
2	712	.44	25	3,818	2.37
3	4,261	2.65	26	1,329	.83
4	7,566	4.70	27	3,093	1.92
5	7,531	4.68	28	3,527	2.19
6	7,931	4.93	29	3,676	2.28
7	7,704	4.79	30	4,303	2.67
8	7,873	4.89	31	3,429	2.13
9	1,400	.87	32	2,437	1.51
10	7,012	4.36	33	1,144	.71
11	7,795	4.84	34	4,155	2.58
12	5,095	3.17	35	4,538	2.82
13	6,913	4.30	36	3,701	2.30
14	4,673	2.90	37	3,000	1.86
15	3,283	2.04	38	2,592	1.61
16	1,657	1.03	39	2,778	1.73
17	4,340	2.70	40	2,056	1.28
18	3,875	2.41	41	3,328	2.07
19	4,210	2.62	42	4,449	2.77
20	3,970	2.47	43	3,562	2.21
21	3,431	2.13	44	3,690	2.29
22	1,992	1.24	45	4,750	2.95
23	1,610	1.00	46	4,759	2.96

Distance between monuments

Monument No.	Distance		Monument No.	Distance	
	Meters	Miles		Meters	Miles
47	4,774	2.97	85	3,983	2.48
48	3,889	2.42	86	4,444	2.76
49	4,443	2.76	87	6,326	3.93
50	4,382	2.72	88	6,685	4.15
51	4,049	2.52	89	3,463	2.15
52	3,321	2.06	90	3,991	2.48
53	542	.34	91	5,148	3.20
54	3,981	2.47	92	5,693	3.54
55	6,768	4.21	93	4,463	2.77
56	2,138	1.33	94	3,309	2.06
57	3,617	2.25	95	3,475	2.16
58	5,610	3.49	96	3,074	1.91
59	2,093	1.30	97	3,624	2.25
60	6,477	4.03	98	3,345	2.08
61	5,775	3.59	99	4,993	3.10
62	2,190	1.36	100	4,858	3.02
63	3,334	2.07	101	442	.27
64	4,012	2.49	102	1,729	1.07
65	6,176	3.84	103	5,115	3.18
66	5,404	3.36	104	2,353	1.46
67	4,808	2.99	105	2,789	1.73
68	3,314	2.06	106	6,481	4.03
69	4,985	3.10	107	2,928	1.82
70	3,324	2.07	108	1,527	.95
71	6,090	3.78	109	3,557	2.21
72	1,127	.70	110	3,328	2.07
73	2,531	1.57	111	2,020	1.26
74	3,793	2.36	112	2,679	1.67
75	5,679	3.53	113	6,114	3.80
76	4,034	2.51	114	1,992	1.24
77	3,483	2.16	115	3,046	1.89
78	1,967	1.22	116	3,862	2.40
79	5,988	3.72	117	4,143	2.57
80	4,678	2.91	118	1,865	1.16
81	1,328	.83	119	5,142	3.20
82	1,796	1.12	120	2,471	1.54
83	1,870	1.16	121	1,076	.67
84	6,060	3.77	122	2,257	.16

Distance between monuments

Monument No.	Distance		Monument No.	Distance	
	Meters	Miles		Meters	Miles
123	1,920	1.19	161	1,989	1.24
124	3,063	1.90	162	5,289	3.29
125	3,066	1.91	163	1,968	1.22
126	4,404	2.74	164	5,201	3.23
127	232	.14	165	6,085	3.78
128	394	.24	166	2,944	1.83
129	5,642	3.51	167	4,753	2.95
130	3,663	2.28	168	2,479	1.54
131	5,692	3.54	169	5,051	3.14
132	5,348	3.32	170	4,464	2.77
133	2,531	1.57	171	3,368	2.09
134	3,050	1.90	172	4,471	2.79
135	6,179	3.84	173	5,778	3.59
136	2,532	1.57	174	3,272	2.03
137	4,239	2.63	175	4,928	3.06
138	4,567	2.84	176	4,286	2.66
139	3,592	2.23	177	6,011	3.74
140	4,569	2.84	178	7,692	4.78
141	2,029	1.26	179	4,698	2.92
142	5,522	3.43	180	5,701	3.54
143	5,632	3.50	181	7,711	4.79
144	4,534	2.82	182	7,296	4.53
145	5,527	3.44	183	7,903	4.91
146	6,370	3.96	184	4,213	2.62
147	7,281	4.53	185	4,496	2.79
148	3,888	2.42	186	4,505	2.80
149	4,661	2.90	187	4,005	2.49
150	4,058	2.52	188	7,823	4.86
151	3,606	2.24	189	6,272	3.90
152	5,237	3.25	190	3,551	2.21
153	4,564	2.84	191	3,255	2.02
154	5,984	3.72	192	3,139	1.95
155	5,294	3.29	193	5,378	3.34
156	5,384	3.35	194	5,725	3.56
157	4,666	2.90	195	3,126	1.94
158	3,953	2.46	196	5,938	3.69
159	3,840	2.39	197	7,235	4.50
160	4,066	2.53	198	7,775	4.83

Appendix A

Distance between monuments

Monument No.	Distance	
	Meters	Miles
199	7,595	4.72
200	7,561	4.70
201	6,409	3.99
202	4,734	2.94
203	4,668	2.90
204	4,190	2.60
205	3,000	1.86

Appendix B

Common and Scientific Plant Names

Common names	*Scientific names*
Acacia, fern	*Acacia angustissima* (Miller) Kuntze
Acacia, whitethorn	*A. constricta* Benth.
Agave, Palmer	*Agave palmeri* Engelm.
Agave, American	*A. americana* L.
Agave, Schott	*A. schottii* Engelm.
Algerita	*Berberis haematocarpa* Wooton
Amole	*Agave schottii* Engelm.
Amoreuxia	*Amoreuxia palmatifida* Moc. & Sesse
Arrow weed	*Pluchea sericea* (Nutt.) Cov.
Ash, velvet	*Fraxinus velutina* Torrey
Batamote	*Baccharis sarothroides* A. Gray
Beardgrass, cane	*Andropogon barbinodis* Lag.
Beardgrass, Texas	*A. cirratus* Hack.
Beargrass	*Nolina microcarpa* Wats.
Bee brush	*Lippia sp.*
Beebrush, Wright	*L. wrightii* A. Gray
Bird of paradise	*Caesalpinia gilliesii* Wall.
Bladderpod	*Lesquerella fendleri* (Gray) Wats.
Bluestem, Texas	*Andropogon cirratus* Hack.
Borage	*Boraginaceae*
Bristlegrass, plains	*Setaria macrostachya* H.B.K.
Brittlebush	*Encelia farinosa* Gray
Broom, turpentine	*Thamnosma texana* (Gray) Torr.
Buckwheat	*Eriogonum sp.*

Buckwheat, shrubby	*E. fasciculatum* Benth.
Buffalograss	*Buchloe dactyloides* (Nutt.) Engelm.
Bullgrass	*Muhlenbergia emersleyi* Vasey
Bunchgrass, wooly	*Elyonurus barbiculmis* Hack.
Burroweed	*Haplopappus tenuisectus* (Greene) Blake
Burroweed, spiny	*H. spinulosus* (Pursh) DC.
Bursage, triangle	*Ambrosia deltoidea* (Torrey) Payne
Bursage, white	*A. dumosa* (A. Gray) Payne
Cactus, beavertail	*Opuntia basilaris* Engelm. & Bigel.
Cactus, hedgehog	*Echinocereus engelmannii* (Parry) Lemaire
Cactus, organpipe	*Cereus thurberi* (Engelm) Britt. & Rose
Canaigre	*Rumex hymenosepalus* Torr.
Casuarina	*Casuarina sp.*
Catclaw	*Acacia greggii* A. Gray
Century plant	*Agave americana* L.
	A. palmeri Engelm.
Chicalote	*Argemone sp.*
Chittamwood	*Bumelia lanuginosa* (Michx.) Pers. var. *rigida* A. Gray
Cholla	*Opuntia sp.*
Cholla, buckhorn	*O. acanthocarpa* Engelm. & Bigel.
Cholla, cane	*O. spinosior* Engelm. & Bigel.
Cholla, diamond	*O. ramosissima* Engelm.
Cholla, jumping	*O. fulgida* Engelm.
Cholla, pencil	*O. arbuscula* Engelm.
Cholla, teddybear	*O. bigelovii* Engelm.
Chuparosa	*Justicia californica* (Benth.) D. Gibson
Coldenia, gray	*Coldenia canescens* DC.
Coldenia, Palmer	*C. palmeri* Gray
Combseed	*Pectocarya sp.*
Copal	*Bursera microphylla* Gray
Cottongrass, Arizona	*Trichachne californica* (Benth.) Chase
Cottonwood, Fremont	*Populus fremontii* Wats.
Creosotebush	*Larrea tridentata* (DC.) Coville
Croton	*Croton corymbulosus* Engelm.
Croton, Sonoran	*C. sonorae* Torr.
Crownbeard	*Verbesina encelioides* (Cav.) Benth. & Hook
Crucifixion thorn	*Koeberlinia spinosa* Zucc.
Crucillo	*Condalia warnockii* M. C. Johnston var. *Kearneyana* M. C. Johnston

Curly-mesquite	*Hilaria belangeri* (Steud.) Nash
Dalea, feather	*Dalea formosa* Torr.
Dalea, Gregg	*D. pulchra* Gentry
Dalea, Parry	*D. parryi* Torr. & Gray
Deergrass	*Muhlenbergia rigens* (Benth.) Hitchc.
Desert broom	*Baccharis sarothroides* Gray
Ditaxis	*Ditaxis sp.*
Dropseed, sand	*Sporobolus cryptandrus* (Torr.) Gray
Fagonia	*Fagonia californica* Benth.
False-mesquite	*Calliandra eriophylla* Benth.
Fiddleneck	*Amsinckia sp.*
Filaree	*Erodium cicutarium* (L.) L'Hér.
Filaree, Texas	*E. texanum* Gray
Fluffgrass	*Tridens pulchellus* (H.B.K.) Hitchc.
Galletagrass	*Hilaria jamesii* (Torr.) Benth.
Golden-eye	*Viguiera sp.*
Golden-eye, annual	*V. annua* (Jones) Blake
Grama, black	*Bouteloua eriopoda* Torr.
Grama, blue	*B. gracilis* (H.B.K.) Lag.
Grama, hairy	*B. hirsuta* Lag.
Grama, purple	*B. radicosa* (Fourn.) Griffiths
Grama, Rothrock	*B. rothrockii* Vasey
Grama, sideoats	*B. curtipendula* (Michx.) Torr.
Grama, six-week	*B. aristidoides* (H.B.K.) Griseb.
Grama, slender	*B. filiformis* (Fourn.) Griffiths
Grama, sprucetop	*B. chondrosioides* (H.B.K.) Benth.
Granejo	*Celtis pallida* Torr.
Groundsel	*Senecio sp.*
Groundsel, threadleaf	*S. longilobis* Benth.
Hackberry, desert	*Celtis pallida* Torr.
Hackberry, western	*C. reticulata* Torr.
Hierba de la flecha	*Sapium biloculare* (Wats.) Pax.
Hind mariola	*Solanum hindsiana* Benth.
Holly, desert	*Perezia nana* Gray
Holly, Wright desert	*P. wrightii* Gray
Hopbush	*Dodonaea viscosa* Jacq.
Huajillo	*Calliandra eriophylla* Benth.
Indigobush, Gregg	*Dalea pulchra* Gentry
Indigobush Parry	*D. parryi* Torr. & Gray

Ipomoea	*Ipomoea longifolia* Benth.
Ironwood	*Olneya tesota* Gray
Jojoba	*Simmondsia californica* Nutt.
Juniper, alligator	*Juniperus deppeana* Steud.
Juniper, one-seed	*J. monosperma* (Engelm.) Sarg.
Kidneywood	*Eysenhardtia polystachya* (Ortega) Sarg.
Lavender, desert	*Hyptis emoryi* Torr.
Lemonade-berry	*Rhus trilobata* (Nutt.)
Limberbush	*Jatropha cardiophylla* (Torrey) Muehl. Arg.
Limberbush, wedgeleaf	*J. cuneata* Wiggins & Rollins
Locoweed	*Astragalus sp.*
Lovegrass, Boer	*Eragrostis chloromelas* Steud.
Lovegrass, Lehmann	*E. lehmanniana* Nees.
Lovegrass, plains	*E. intermedia* Hitchc.
Lupine	*Lupinus sp.*
Mallow	*Sphaeralcea sp.*
Manzanita, Mexican	*Arctostaphylos pungens* H.B.K.
Mariola	*Parthenium incanum* H.B.K.
Menodora	*Menodora scabra* Gray
Mercury, Pringle three-seeded	*Acalypha pringlei* Wats.
Mescal	*Agave palmeri* Engelm.
Mesquite, honey	*Prosopis juliflora* (Swartz) DC. var. *glandulosa* (Torrey) Cockerell
Mesquite, western honey	*P. juliflora* (Swartz) DC. var. *torreyana* L. Benson
Mesquite, velvet	*P. juliflora* (Swartz) DC. var. *velutina* (Weston) Sarg.
Mimosa, velvet-pod	*Mimosa dysocarpa* Benth.
Morning glory	*Convolvulus sp.*
Mountain-mahogany	*Cerocarpus montanus* Raf. var. *glaber* (S. Wats.) F. L. Martin
Muhly, Arizona	*Muhlenbergia arizonica* Scribn.
Muhly, bush	*M. porteri* Scribn.
Muhly, mountain	*M. montana* (Nutt.) Hitchc.
Mulberry	*Morus rubra* L.
Mustard (family)	*Cruciferae*
Mustard, Thurber	*Lepidium thurberi* Woot

Needlegrass	*Stipa sp.*
Nolina, bigelow	*Nolina bigelovii* (Torr.) Wats.
Oak	*Quercus sp.*
Oak, desert scrub	*Q. turbinella* Greene
Oak, Emory	*Q. emoryi* Torr.
Oak, Mexican blue	*Q. oblongifolia* Torr.
Oak, scrub	*Q. pungens* Liebm.
Oak, silverleaf	*Q. hypoleucoides* Camus
Oak, Toumey	*Q. toumeyi* Sarg.
Ocotillo	*Fouquieria splendens* Engelm.
Odora	*Porophyllum gracile* Benth.
Palmilla	*Yucca elata* Engelm.
Paloverde, blue	*Cercidium floridum* Benth.
Paloverde, foothill	*C. microphyllum* (Torr.) Rose & Johnston
Penstemon	*Penstemon parryi* Gray
Peppergrass	*Lepidium sp.*
Pinyon, Mexican	*Pinus cembroides* Zucc.
Plantain	*Plantago sp.*
Prickle-poppy	*Argemone sp.*
Pricklypear	*Opuntia sp.*
Pricklypear, Engelmann	*O. phaeacantha* Engelm.
Pricklypear, purple	*O. violacea* Engelm.
Ragweed	*Ambrosia psilostachya* DC.
Ratany, range	*Krameria sp.*
Rose-mallow	*Hibiscus sp.*
Rosewood, Arizona	*Vauquelinia californica* (Torr.) Sarg.
Sacaton	*Sporobolus wrightii* Munro
Sage, Parry	*Salvia parryi* Gray
Sagebrush, Louisiana	*Artemisia ludoviciana* Nutt.
Sagebrush, threadleaf	*A. filifolia* Torr.
Sagebrush, wooly	*A. ludoviciana* Nutt.
Sahuaro	*Cereus giganteus* (Engelm.) Britt. & Rose
Saltbush, desert	*Atriplex polycarpa* (Torr.) Wats.
Saltbush, fourwing	*A. canescens* (Pursh.) Nutt.
Saltbush threadleaf	*A. linearis* Wats.
Schismus, six-week	*Schismus barbatus* (L) Thell.

Sedge	*Carex sp.*
Seepweed	*Suaeda torreyana* Wats.
Senna shrubby	*Cassia wislizeni* Gray
Silktassel, bush	*Garrya wrightii* Torr.
Snakeweed	*Gutierrezia sarothrae* (Pursh.) Britt. & Rusby
Sotol	*Dasylirion wheeleri* Wats.
Sprangletop	*Leptochloa dubia* (H.B.K.) Nees.
Squawbush	*Rhus choriophylla* Woot. & Standl.
Squawbush, littleleaf	*R. microphylla* Engelm.
Squirreltail	*Sitanion hystrix* (Nutt.) J. G. Smith
Sumac, evergreen	*Rhus choriophylla* Woot. & Standl.
Sunflower (family)	*Compositae*
Sycamore, Arizona	*Platanus wrightii* Wats.
Tamarisk	*Tamarix sp.*
Tanglehead	*Heteropogon contortus* (L.) Beauv.
Tarbush	*Flourensia cernua* DC.
Tea, Mormon	*Ephedra trifurca* Torr.
Thistle, Russian	*Salsola kali* L.
Three-awn	*Aristida spp.*
Three-awn, California	*A. californica* Thurb.
Three-awn, red	*A. longiseta* Steud.
Three-awn, Santa Rita	*A. glabrata* (Vasey) Hitchc.
Three-awn, six-week	*A. adscensionis* L.
Tobosa, tobosagrass	*Hilaria mutica* (Buchl.) Benth.
Tobosagrass, big	*H. rigida* (Thurb.) Benth.
Tomatillo	*Lycium sp.*
Torote	*Bursera microphylla* Gray
Tridens, large-flowered	*Tridens grandiflorus* (Vasey) Woot. & Standl.
Tridens, slim	*T. muticus* (Torr.) Nash.
Trixis	*Trixis californica* Kell.
Turpentine bush	*Haplopappus laricifolius* A. Gray
Twinflower, southwestern	*Janusia gracilis* Gray.
Vetch, deer	*Lotus sp.*
Wait-a-minute bush	*Mimosa biuncifera* Benth.
Walnut, Arizona	*Juglans major* (Torr.) Heller
Whitethorn	*Acacia constricta* Benth.
Whitethorn, spring	*A. vernicosa* Standl.
Willow, black	*Salix gooddingii* Ball

Winterfat	*Eurotia lanata* (Pursh) Moq.
Wolfspike	*Elyonurus barbiculmis* Hack.
Wolftail	*Lycurus phleoides* H.B.K.
Yerba de pasmo	*Baccharis pteronioides* DC.
Yucca, banana	*Yucca baccata* Torr.
Yucca, mountain	*Y. schottii* Engelm.
Zinnia, desert	*Zinnia pumila* Gray
Zinnia, large-flowered	*Z. grandiflora* Nutt.

Pertinent Literature

Arnold, Joseph F. 1963. Use of fire in the management of Arizona watersheds. *Proc. 2nd Ann. Tall Timbers Fire Ecology Conference:* 99–111. Tallahassee, Fla.

Bahre, Conrad J. 1985. Wildfire in southeastern Arizona between 1859 and 1890. *Desert Plants* 7(4):190–94.

Bancroft, H. H. n.d. (1890?). *History of Arizona and New Mexico, II.* New York: The Bancroft Co.

Bartlett, John Russell. 1854. *Personal narrative of explorations and incidents in Texas, New Mexico, California, Sonora and Chihuahua.* 2 vols. New York: D. Appleton & Co.

Bell, James G. 1932. A log of the Texas-California cattle trail, 1854. *Southwestern Historical Quarterly* 35:290–316.

Betancourt, J. L., and R. M. Turner. 1985. History of the Santa Cruz River. *In preparation.* U.S. Geological Survey, Tucson.

Biswell, Harold H. 1974. Effects of fire on chaparral. *In* T. T. Kozlowski and C. E. Ahlgren, eds., *Fire and Ecosystems.* New York, San Francisco, London: Academic Press.

Bolton, Herbert E. 1932. *The padre on horseback.* San Francisco: Sonora Press.

———. 1948. *Kino's historical memoirs of Pimeria Alta.* 2 vols. Berkeley and Los Angeles: Univ. Calif. Press.

Box, Michael James. 1869. *Adventures and explorations in New and Old Mexico.* New York: James Miller.

Brown, David E., ed. 1982. Biotic Communities of the American Southwest. *Desert Plants* 4(1–4):3–341.

Brown, David E., Charles H. Lowe, and Charles P. Pase. 1979. A digitized classification system for the biotic communities of North America, with community (series) and association examples for the Southwest. *Jour. AZ-NE Acad. Sci.* 14 (supplement 1): 1–16.

Literature Cited

Bryan, Kirk. 1925. Date of channel trenching (arroyo cutting) in the arid southwest. *Science* (N.S.) 62:338–44.

Bufkin, D. H. 1983. The making of a boundary between the United States and Mexico. *The Cochise Quarterly* (1,2):3–30.

Cable, Dwight R. 1967. Fire effects on semidesert grasses and shrubs. *Jour. Range Mgt.* 20 (3):170–76.

Cooke, P. St. George. 1849. *Report of the Secretary of War communicating . . . a copy of the official journal of Lieutenant Colonel Philip St. George Cooke, from Santa Fe to San Diego,* etc. *In* U.S. Senate Public Documents No. 2. 31st Congress, Special Session.

———. 1938. *Exploring Southwestern Trails, 1846–1854.* Ed. by Ralph P. Bieber. Glendale, Calif: The Arthur H. Clark Company.

Cooke, Ronald V., and R. W. Reeves. 1976. *Arroyos and environmental change in the American Southwest.* Oxford: Clarendon Press.

Culley, M. J. 1943. Grass grows in summer or not at all. *Am. Hereford Jour.* 34:8–10.

Denevan, William M. 1967. Livestock numbers in nineteenth-century New Mexico and the problem of gullying in the Southwest. *Ann. Assoc. Am. Geogr.* 57:691–703.

Emory, William H. 1848. *Notes of a Military Reconnaissance, from Fort Leavenworth, in Missouri, to San Diego in California.* 614 pp, plates, maps. U.S. Cong. House Ex. Doc. 41. Washington, D.C.: Wendell & Van Benthuysen.

———. 1857. *Report on the United States and Mexican Boundary Survey made under the direction of the Secretary of the Interior.* Vol. I. Washington, D.C.: A. O. P. Nicholson.

Evans, George W. B. 1945. *Mexican gold trail: the Journal of a Forty-niner.* Ed. by Glenn S. Dumke. San Marino: Huntington Library.

Faulk, Odie B. 1967. *Too Far North—Too Far South.* Los Angeles: Westernlore Press.

Ford, M. J. 1982. *The Changing Climate: Responses of the natural fauna and flora.* London: George Allen and Unwin Press.

Haskett, B. 1935. Early history of the cattle industry in Arizona. *AZ Hist. Rev.* 6:3–42.

Hastings, J. R. Vegetation change and arroyo cutting in southeastern Arizona. *Jour. AZ Acad. Sci.* 1(2):60–67.

Hastings, R. H., and R. M. Turner. 1956. *The Changing Mile: An ecological study of vegetation change with time in the lower mile of an arid and semidesert region.* Tucson: Univ. Ariz. Press.

Helms, Christopher, ed. 1980. *The Sonoran Desert.* Las Vegas: KC Publications.

Hine, Robert V. 1968. *Bartlett's West: Drawing the Mexican Boundary.* New Haven, Conn.: Yale Univ. Press.

Humphrey, R. R. 1958. The desert grassland. *Bot. Rev.* 24:193–252.
———. 1974. Fire in the deserts and desert grassland of North America. *In* T. T. Kozlowski, and C. E. Ahlgren, eds. *Fire and Ecosystems.* New York, San Francisco, London: Academic Press.
———. 1981. The Sonoran Desert. *Cactus and Succulent Journal* (U.S.) 53:246–54.
Jaeger, Edmund C. 1957. *The North American Deserts.* Stanford: Stanford Univ. Press.
Martin, S. Clark. 1973. Invasion of semidesert grassland by velvet mesquite and associated vegetation changes. *Jour. AZ Acad. Sci.* 8:127–34.
McGee, W. J. 1901. The old Yuma trail. *Nat'l Geogr. Mag.* 12:103–7.
Mearns, E. A. 1907. *Mammals of the Mexican Boundary of the United States.* Smithsonian Institution, United States National Museum Bull. 56. Washington, D.C.: Gov't Printing Office.
Minnich, R. A. 1983. Fire mosaics in southern California and northern Baja California. *Science* 219:1287–94.
Powell, M. H. T. 1931. *The Santa Fe Trail to California, 1849–1852.* San Francisco: Grabhorn Press.
Pyne, S. J. 1982. *A cultured history of wildland and rural fire.* Princeton, N.J.: Princeton Univ. Press.
Reynolds, H. G. 1954. Meeting drought on southern Arizona rangelands. *Jour. Range Mgt.* 7:33–40.
Rothrock, J. T. 1875. *Preliminary and general botanical report, with remarks upon the general topography of the region traversed.* Appendix H of Annual Report upon the Geographical Exploration and Surveys West of One Hundredth Meridian (Appendix 100 of the Annual Report of the Chief of Engineers for 1875). Washington, D.C.: Gov't Printing Office.
Sauer, C. O. 1950. Grassland climax, fire and man. *Jour. Range Mgt.* 3:16–21.
Sellers, William D. 1960. Precipitation trends in Arizona and western New Mexico. *Proc. 28th Annual Snow Conference:* 81–94. Santa Fe, N.M.
Senate Document 247. 1898a. United States Boundary Commission. *Report of the Boundary Commission upon the survey and remarking of the boundary between the United States and Mexico west of the Rio Grande, 1891–1896.* Parts I and II, Senate Document 247, 55th Congress, 2nd session. Washington, D.C.: Gov't Printing Office.
Senate Document 247. 1898b. United States Boundary Commission. *Report of the Boundary Commission upon the survey and remarking of the boundary between the United States and Mexico west of the Rio Grande, 1891–1896. ALBUM.* Senate Document 247, 55th Congress, 2nd session. Washington, D.C.: Gov't Printing Office.

Shreve, Forrest. 1917. A map of the vegetation of the United States. *Geog. Rev.* 3:119–25.

———. 1915. *The vegetation of a desert mountain range as conditioned by climatic factors.* Carnegie Institution of Washington Publication 217. Washington, D.C.: Carnegie Institution.

———. 1942. The desert vegetation of North America. *Botanical Review* 8:195–246.

———. 1951. *Vegetation of the Sonoran Desert.* Carnegie Institution of Washington Publication 591. Maps. Washington, D.C.: Carnegie Institution.

Shreve, Forrest, and Ira L. Wiggins. 1964. *Vegetation and flora of the Sonoran Desert.* 2 vols. Stanford: Stanford Univ. Press.

Stewart, O. C. 1954. The forgotten side of ethnogeography. *Reprint* from *Method and Perspective in Anthropology,* pp. 221–310. Minneapolis: Univ. Minn. Press.

———. 1955. Why were the prairies treeless? *Southwestern Lore* 20:59–64.

Van Devender, Thomas R., and W. G. Spaulding. 1979. Development of vegetation and climate in the Southwestern United States. *Science:* 24:701–10.

Von Eschen, G. F. 1958. Climatic trends in New Mexico. *Weatherwise* 11:191–95.

Wagoner, Jay J. 1952. *History of the Cattle Industry in Southern Arizona, 1540–1940.* Univ. of Ariz. Social Science Bulletin No. 20. Tucson: Univ. of Ariz.

———. 1975. *Early Arizona.* Tucson: Univ. Ariz. Press.